Chris Lamoureux

Lecture Notes in Statistics

Vol. 1: R. A. Fisher: An Appreciation. Edited by S. E. Fienberg and D. V. Hinkley. xi, 208 pages, 1980.

Vol. 2: Mathematical Statistics and Probability Theory. Proceedings 1978. Edited by W. Klonecki, A. Kozek, and J. Rosiński. xxiv, 373 pages, 1980.

Vol. 3: B. D. Spencer, Benefit-Cost Analysis of Data Used to Allocate Funds. viii, 296 pages, 1980.

Vol. 4: E. A. van Doorn, Stochastic Monotonicity and Queueing Applications of Birth-Death Processes. vi, 118 pages, 1981.

Vol. 5: T. Rolski, Stationary Random Processes Associated with Point Processes. vi, 139 pages, 1981.

Vol. 6: S. S. Gupta and D.-Y. Huang, Multiple Statistical Decision Theory: Recent Developments. viii, 104 pages, 1981.

Vol. 7: M. Akahira and K. Takeuchi, Asymptotic Efficiency of Statistical Estimators. viii, 242 pages, 1981.

Vol. 8: The First Pannonian Symposium on Mathematical Statistics. Edited by P. Révész, L. Schmetterer, and V. M. Zolotarev. vi, 308 pages, 1981.

Vol. 9: B. Jørgensen, Statistical Properties of the Generalized Inverse Gaussian Distribution. vi, 188 pages, 1981.

Vol. 10: A. A. McIntosh, Fitting Linear Models: An Application of Conjugate Gradient Algorithms. vi, 200 pages, 1982.

Vol. 11: D. F. Nicholls and B. G. Quinn, Random Coefficient Autoregressive Models: An Introduction. v, 154 pages, 1982.

Vol. 12: M. Jacobsen, Statistical Analysis of Counting Processes. vii, 226 pages, 1982.

Vol. 13: J. Pfanzagl (with the assistance of W. Wefelmeyer), Contributions to a General Asymptotic Statistical Theory. vii, 315 pages, 1982.

Vol. 14: GLIM 82: Proceedings of the International Conference on Generalised Linear Models. Edited by R. Gilchrist. v, 188 pages, 1982.

Vol. 15: K. R. W. Brewer and M. Hanif, Sampling with Unequal Probabilities. ix, 164 pages, 1983.

Vol. 16: Specifying Statistical Models: From Parametric to Non-Parametric, Using Bayesian or Non-Bayesian Approaches. Edited by J. P. Florens, M. Mouchart, J. P. Raoult, L. Simar, and A. F. M. Smith. xi, 204 pages, 1983.

Vol. 17: I. V. Basawa and D. J. Scott, Asymptotic Optimal Inference for Non-Ergodic Models. ix, 170 pages, 1983.

Continued

Lecture Notes in Statistics

Edited by D. Brillinger, S. Fienberg, J. Gani,
J. Hartigan, and K. Krickeberg

19

Luisa Turrin Fernholz

von Mises Calculus for Statistical Functionals

Springer-Verlag
New York Berlin Heidelberg Tokyo

Luisa Turrin Fernholz
Department of Statistics
Princeton University
Fine Hall, P.O. Box 37
Princeton, NJ 08544
U.S.A.

AMS Subject Classifications: 62A99, 62E20

Library of Congress Cataloging in Publication Data
Fernholz, Luisa Turrin
    Von Mises calculus for statistical functionals.
    (Lecture notes in statistics ; 19)
    Bibliography: p.
    1. Asymptotic distribution (Probability theory)
2. Statistical functionals.   3. Estimation theory.
I. Title.   II. Series: Lecture notes in statistics
(Springer-Verlag) ; v. 19.
QA276.7.F47   1983   519.5   83-12524

With one illustration

Printed and bound by R.R. Donnelley & Sons, Harrisonburg, VA.
Printed in the United States of America.

9 8 7 6 5 4 3 2 1

ISBN 0-387-90899-4 Springer-Verlag New York Berlin Heidelberg Tokyo
ISBN 3-540-90899-4 Springer-Verlag Berlin Heidelberg New York Tokyo

PREFACE

About forty years ago, Richard von Mises proposed a theory for the analysis of the asymptotic behavior of nonlinear statistical functionals based on the differentiability properties of these functionals. His theory was largely neglected until the late 1960's when it experienced a renaissance due to developments in the field of robust statistics. In particular, the "Volterra" derivative used by von Mises evolved into the influence curve, which was used to provide information about the sensitivity of an estimator to outliers, as well as the estimator's asymptotic variance. Moreover, with the "Princeton Robustness Study" (Andrews et al. (1972)), there began a proliferation of new robust statistics, and the formal von Mises calculations provided a convenient heuristic tool for the analysis of the asymptotic distributions of these statistics. In the last few years, these calculations have been put in a more rigorous setting based on the Fréchet and Hadamard, or compact, derivatives.

The purpose of these notes is to provide von Mises' theory with a rigorous mathematical framework which is sufficiently straightforward so that it can be applied routinely with little more effort than is required for the calculation of the influence curve. The approach presented here is based on the Hadamard derivative and is applicable to diverse forms of statistical functionals.

This work is partially derived from the first part of my doctoral dissertation, submitted in 1979 to Rutgers University, and I wish to thank my advisor, Professor Robert H. Berk, for introducing me to this topic and for his guidance and helpful suggestions. I thank my husband, Bob, whose encouragement and assistance were vital to the successful completion of this work. I thank my earlier mathematics teachers at Universidad de Buenos Aires, especially Dr. Manuel Balanzat, from whom I acquired the basic mathematical background which made this research possible. I thank the Statistics Department of Princeton University for its support during the preparation of these notes. Finally, I thank Ms. Maureen Kirkham for typing the manuscript.

Luisa Turrin Fernholz
Princeton, New Jersey
April, 1983

TABLE OF CONTENTS

CHAPTER I

INTRODUCTION

A statistic can frequently be considered as a functional on a space
of distribution functions. Often such a statistical functional possesses
differentiability properties which provide information about its asymptotic behavior. These basic ideas were introduced by R. von Mises
(1947), who developed a theory for the analysis of the asymptotic distribution of statistical functionals, using a form of Taylor expansion involving the derivatives of the functionals.

Von Mises observed that a statistic $T(F_n)$ can be expressed as

$$(1.1) \qquad T(F_n) = T(F) + T_F'(F_n - F) + \text{Rem}(F_n - F) \, ,$$

where F_n is the empirical distribution function corresponding to a
sample $X_1, \ldots, X_n$ from a population with distribution function F ,
$T(F)$ is the parameter to be estimated, and T_F' is the derivative of
the functional T at F . These ideas are presented in Chapter II and
the properties of the different terms of the expansion (1.1) are discussed. In particular, the term $T_F'(F_n - F)$ is linear and is therefore a sum
of independent identically distributed random variables, so the central
limit theorem implies that for some finite $\sigma^2 > 0$,

$$(1.2) \qquad \sqrt{n} \; T_F'(F_n - F) \xrightarrow{\;\mathcal{D}\;} N(0, \sigma^2) \, .$$

1

Under appropriate conditions, the remainder term satisfies

$$(1.3) \qquad \sqrt{n}\ \mathrm{Rem}(F_n - F) \xrightarrow{P} 0\ .$$

If (1.2) and (1.3) hold, then it follows that

$$(1.4) \qquad \sqrt{n}(T(F_n) - T(F)) \xrightarrow{D} N(0, \sigma^2)\ .$$

Much of the work that has followed von Mises' original contribution has been characterized by the use of a different form of derivative for each distinct type of statistical functional T . In these notes, we present a unified approach to von Mises' theory which can be applied uniformly to diverse classes of statistics. We first observe that a statistical functional induces a functional on the space $D[0,1]$ of right continuous functions with left limits. Then, using the Hadamard, or compact, derivative, we develop a calculus for functionals defined on $D[0,1]$. This calculus is used to establish the differentiability of a given statistical functional, after which an expansion of the form (1.1) can be generated and the asymptotic result (1.4) obtained.

Since the expansion (1.1) is based on some form of differentiation, in Chapter III we consider three distinct types of derivative: Gateaux, Hadamard, and Fréchet. The derivative originally used by von Mises was similar to, but not precisely the same as, the Gateaux derivative. To prove the validity of condition (1.3), von Mises assumed the existence of the second order derivative, but this rather strong condition is seldom satisfied. The slightly weaker assumption of (single) Fréchet differentiability implies (1.3), but still this is too strong a requirement since few statistical functionals are Fréchet differentiable. Hadamard differentiability is a weaker condition than Fréchet differentiability, and many statistical functionals can be shown to be Hadamard differentiable. This form of derivative, which was first used in statistics by J. Reeds (1976), also implies condition (1.3).

In Chapter IV we review some probability theory on $C[0,1]$, the space of continuous functions on $[0,1]$, and on $D[0,1]$, including Reeds' (1976) work relating the Hadamard derivative to the convergence of (1.3). The main result of Chapter IV is that if a statistical functional induces a Hadamard differentiable functional on $D[0,1]$ or $C[0,1]$, then the statistical functional is asymptotically normal, as in (1.4).

The three main classes of robust estimators, M-, L-, and R-estimators, are introduced in Chapter V. Certain properties related to the Hadamard differentiability of these functionals are established, to be applied later in Chapter VII.

In Chapter VI we develop some basic elements of a differential calculus on function spaces. We present a number of theorems on the Hadamard differentiability of transformations on $D[0,1]$, including inversion, composition, and smoothing. We also prove an implicit function theorem which can be applied to statistical functionals that are defined implicitly in the form of a root $T(F_n) = \theta$ of an equation

$$\Phi(F_n,\theta) = 0 \; ,$$

where Φ is a function of a distribution function F_n and a parameter θ . The implicit function theorem shows that the Hadamard differentiability of Φ implies the Hadamard differentiability of T , when certain conditions are satisfied.

In Chapter VII we apply the von Mises calculus to M-, L-, and R-estimators to derive the asymptotic normality condition (1.4) for these classes of statistics. We also consider sample quantiles, which are shown to induce functionals on $C[0,1]$, and for which we use a parallel and simpler approach. Other applications are also presented.

Finally, in Chapter VIII we show that the Hadamard derivative can be used to study the asymptotic efficiency of a statistical functional.

We follow the approach of Huber (1977) and prove that under certain regularity conditions, a Fisher consistent estimator which is Hadamard differentiable is asymptotically efficient if and only if its influence curve satisfies a relation involving the Fisher score function. This is applied to M-, L-, and R-estimators, as well as the sample median.

It is worthwhile to mention here a few conventions which we shall follow throughout these notes. A function $f: \mathbb{R} \longrightarrow \mathbb{R}$ is <u>nondecreasing</u> if $f(x) \leq f(y)$ for $x \leq y$, and is <u>increasing</u> if $f(x) < f(y)$ for $x < y$; the terms "increasing", "monotone increasing", and "strictly increasing" will all have the same meaning and will be used interchangeably. Analogously for "nonincreasing" and "decreasing". The functional notation $f: A \longrightarrow B$ will sometimes be used even when the domain of f is a proper subset of A . The generalized inverse of a function $G: [a,b] \longrightarrow \mathbb{R}$ is defined by

$$G^{-1}(x) = \inf \{b,t: G(t) \geq x\} ,$$

which coincides with the usual inverse for continuous, increasing functions.

A list of symbols and abbreviations is provided at the end of these notes.

CHAPTER II

VON MISES' METHOD

In this chapter we present the general structure of von Mises'
approach to the analysis of the asymptotic behavior of statistical functionals. The basic technique was introduced by von Mises (1947) and has
been extended in various directions by several authors: Filippova
(1962), Reeds (1976), Huber (1977,1981), and Serfling (1980). One
result of these extensions is that the field has become divergent, with
ad hoc techniques applied in different situations. In the chapters that
follow we shall try to establish a unified methodology that can be
applied to wide classes of statistics.

Let us now review some of the past work and set the stage for the
later chapters.

2.1 Statistical functionals

Let $X_1,\ldots,X_n$ be a sample from a population with distribution
function (d.f.) F and let $T_n = T_n(X_1,\ldots,X_n)$ be a statistic. If T_n
can be written as a functional T of the empirical d.f. F_n ,
$T_n = T(F_n)$, where T does not depend on n , then T will be called a
statistical functional. The domain of definition of T is assumed to
contain the empirical d.f.'s F_n for all $n \geq 1$, as well as the population d.f. F . Unless otherwise specified, the range of T will be the

set of real numbers. The parameter to be estimated in estimation problems is $T(F)$.

Statistical functionals were introduced by von Mises (1936, 1937, 1947), and are currently used in the theory of robust estimation. The following are some examples of statistical functionals.

Example 2.1.1. Let ϕ be a real valued function and let

$$T_n(X_1,\ldots,X_n) = \frac{1}{n} \sum_{i=1}^{n} \phi(X_i) \ .$$

Then for a general d.f. G , the functional defined by

$$T(G) = \int \phi(x) dG(x)$$

satisfies $T_n(X_1,\ldots,X_n) = T(F_n)$. This is perhaps the simplest form of statistical functional. $\triangle$

Example 2.1.2. Let ψ be a real valued function of two variables and let T_n be defined implicitly by

$$\sum_{i=1}^{n} \psi(X_i,T_n) = 0 \ .$$

The corresponding functional is defined as a solution $T(G) = \theta$ of

$$\int \psi(x,\theta) dG(X) = 0 \ .$$

Estimators of this form are called M-estimators. They are discussed at length in Huber (1981) and will be considered in more detail in Chapter V.

$\triangle$

Functionals of the form

$$(2.1) \qquad\qquad T(G) = \int \phi(x) dG(x)$$

are called <u>linear statistical functionals</u> (or simply, linear functionals).

An application of the central limit theorem shows that for a linear

statistical functional T ,

$$(2.2) \qquad\qquad \sqrt{n}(T(F_n)-T(F)) \xrightarrow{\ D\ } N(0,\sigma^2)$$

provided that

$$0 < \int \phi^2(x)\,dF(x)-\left(\int \phi(x)\,dF(x)\right)^2 = \sigma^2 < \infty \ .$$

The central idea behind von Mises' method is to extend this asymptotic

normality result to statistical functionals which are not linear by means

of an approximation by linear functionals.

2.2 Von Mises expansions

Von Mises (1947) proposed that a Taylor expansion could be used to

approximate statistical functionals by statistical functionals of simpler

form, and that this approximation could be applied to obtain results

about their asymptotic distribution. The first term of such an expansion

is linear, and under appropriate conditions if this term is non-vanish-

ing then the statistical functional can be shown to be asymptotically

normal as in (2.2). The existence of such a Taylor expansion depends on

differentiability properties of the statistical functional, so we shall

now introduce the von Mises derivative (in a heuristic manner) and out-

line how the asymptotic normality results are derived.

Definition 2.2.1: Let $X_1,\ldots,X_n$ be a sample and let T be a function-

al on a convex set of d.f.'s containing all empirical d.f.'s and the

population d.f. F . Let G be a point in this convex set. Then the

von Mises derivative T_F' of T at F is defined by

$$T_F'(G-F) = \frac{d}{dt}\,T(F+t(G-F))\Big|_{t=0}$$

if there exists a real valued function ϕ_F (independent of G) such that

$$T_F'(G-F) = \int \phi_F(x) d(G-F)(x) \ .$$

Δ

Higher order derivatives can be defined in a similar manner. The function ϕ_F is uniquely defined up to an additive constant since $d(G-F)$ has total measure zero. We shall normalize by making

$$\int \phi_F(x) dF(x) = 0 \ .$$

The von Mises derivative has been mistakenly referred to in the statistical literature as the "Volterra derivative". It has also been called the "Gateaux derivative", to which it bears a close resemblance as will be shown in Chapter III. A detailed history of these terms can be found in Reeds (1976).

The function ϕ_F is called the <u>influence curve</u> or <u>influence function</u> of T at F . It is usually defined by

$$\phi_F(x) = \frac{d}{dt} T(F+t(\delta_x-F))\big|_{t=0}$$

where δ_x is the d.f. of the point mass one at x . The usual notation for the influence curve is

$$IC(x;F,T) = \phi_F(x) \ .$$

This function has played an important role in the theory of robust estimation, due to work by F. Hampel (1968, 1974) who observed that for large n , $\phi_F(x)$ measures the effect on T_n of a single additional observation with value x . The influence curve also provides the asymptotic variance when T_n is asymptotically normal. The properties of the influence curve are discussed at length in Hampel (1974).

The existence of the influence curve for a statistical functional does not imply that the functional has a von Mises derivative.

Example 2.2.2. Let F_o be a d.f. with density $F_o'(x) > 0$ for all $x \in \mathbb{R}$, and define the functional

$$T(F) = \lim_{x \to \infty} \left[\frac{F(-x)}{F_o(-x)} - \frac{1-F(x)}{1-F_o(x)} \right]$$

(T measures the relative behavior of F with respect to F_o at $\pm\infty$) . The influence curve of T at F_o exists and is identically zero, but for the d.f. G defined by

$$G(x) = \begin{cases} F_o(x) & \text{for } x < 0 \\ \\ 1 & \text{for } x \geq 0 \end{cases}$$

we have

$$\frac{d}{dt} T(F_o + t(G-F_o)) \big|_{t=0} = 1 \ .$$

Therefore the von Mises derivative of T at F_o does not exist. $\quad\Delta$

Von Mises used the fact that the function

$$A(t) = T(F+t(G-F)) \ , \ t \in [0,1] \ ,$$

can be represented by a Taylor expansion at $t = 0$,

$$A(t) = A(0) + A'(0)t + \ldots + \frac{A^{(k)}(0)t^k}{k!} + \text{Rem}_k$$

where Rem_k is a remainder term. This corresponds to an expansion for T which, in the first order, is given by

$$(2.3) \qquad T(G) = T(F) + T_F'(G-F) + \text{Rem}(G-F)$$

when $t = 1$. This, as well as higher order versions, was used by von Mises to obtain asymptotic distribution results.

For $G = F_n$ the expansion in (2.3) becomes

$$(2.4) \qquad T(F_n) = T(F) + T_F'(F_n - F) + \text{Rem}(F_n - F)$$

$$= T(F) + \int \phi_F(x) d(F_n - F)(x) + \text{Rem}(F_n - F)$$

$$= T(F) + \int \phi_F(x) dF_n(x) + \text{Rem}(F_n - F)$$

since $\int \phi_F(x) dF(x) = 0$. Note that $\text{Rem}(F_n - F)$ depends on F as well as $F_n - F$, but when there is no possibility of confusion we shall suppress this variable in order to simplify notation. The expression (2.4) will be called a <u>von Mises expansion</u> of T at F . The linear term of the expansion is

$$\int \phi_F(x) dF_n(x) = \frac{1}{n} \sum_{i=1}^{n} \phi_F(X_i)$$

and therefore

$$(2.5) \qquad \sqrt{n}(T(F_n) - T(F)) = \frac{1}{\sqrt{n}} \sum_{i=1}^{n} \phi_F(X_i) + \sqrt{n} \text{ Rem}(F_n - F) .$$

If

$$(2.6) \qquad 0 < E\phi_F^2(X) = \sigma^2 < \infty$$

and if

$$(2.7) \qquad \sqrt{n} \text{ Rem}(F_n - F) \xrightarrow{P} 0 ,$$

then the central limit theorem and Slutsky's lemma imply that

$$(2.8) \qquad \sqrt{n}(T(F_n) - T(F)) \xrightarrow{D} N(0, \sigma^2)$$

as $n \longrightarrow \infty$.

Condition (2.6) can be verified immediately by using the influence curve to calculate the asymptotic variance. Condition (2.7) is more complicated and indeed is not satisfied by all statistical functionals that have a von Mises derivative.

Example 2.2.3. Let F be a d.f. on $[0,1]$ and define the statistical functional

$$(2.9) \qquad T(F) = \sum_{x \in [0,1]} (F(x)-F(x^-))^{\alpha}$$

where $F(x^-) = \lim_{x' \uparrow x} F(x')$, and α is a positive number. T measures the jumps of F , and since there are at most a countable number of them, the sum in (2.9) is well defined.

For $\alpha > 1$ and U the uniform distribution on $[0,1]$, we have

$$T(U+t(F-U)) = \sum_x t^{\alpha}(F(x)-F(x^-))^{\alpha}$$

$$= t^{\alpha} \sum_x (F(x)-F(x^-))^{\alpha} \quad ,$$

so

$$\frac{d}{dt} T(U+t(F-U))\big|_{t=0} = 0 \ .$$

Now suppose F_n is the empirical d.f. for U corresponding to a sample of size n . F_n will almost surely have n jumps of height $\frac{1}{n}$, so

$$T(F_n) = n^{1-\alpha} \ .$$

Now

$$\sqrt{n}(T(F_n)-T(U)) = n^{\frac{3}{2}-\alpha}$$

almost surely, so for $1 < \alpha < 3/2$ we see that (2.7) will fail and the asymptotic normality of (2.8) will not hold. Δ

To ensure the validity of condition (2.7) von Mises assumed that $\text{Rem}(F_n - F)$ consisted of a second derivative plus a higher-order remainder term. This assumption that the statistical functional be twice von Mises differentiable is unnecessarily restrictive, but nevertheless was also used in the more recent work of Filippova (1962). Other authors have chosen to use stronger definitions of derivative, for example Kallianpur and Rao (1955), Huber (1981), Boos (1979), and Boos and Serfling (1980) used the Fréchet derivative, and Reeds (1976) used the Hadamard (or compact) derivative. With these stronger forms of differentiation it can usually be shown that (2.7) holds with the existence of the first derivative alone.

2.3 Fréchet derivatives

The usual definition of Fréchet differentiation in a normed vector space is as follows:

<u>Definition 2.3.1</u>. Let V be a normed vector space and let $T: V \longrightarrow \mathbb{R}$ be a function. T is <u>Fréchet differentiable</u> at $F \in V$ if there exists a linear functional $T'_F: V \longrightarrow \mathbb{R}$ such that for all $G \in V$,

$$(2.10) \qquad \lim_{G \to T} \frac{|T(G) - T(F) - T'_F(G-F)|}{\|G-F\|} = 0 \ . \qquad \Delta$$

The linear functional T'_F is called the <u>Fréchet derivative</u> of T at F . Note that it is not required that the linear functional T'_F be continuous, however when T is continuous then so is T'_F (see Dieudonne (1960)).

If we consider d.f.'s on $\mathbb{R}$ to be elements of the vector space of bounded real valued functions and if we equip this space with the uniform topology, i.e. the topology generated by the norm

$$\|G\| = \sup_{x \in \mathbb{R}} |G(x)| \ ,$$

then the existence of the Fréchet derivative for a functional is sufficient to imply that the asymptotic normality condition (2.8) is valid for that statistical functional. An application of the well known properties of the Kolmogorov-Smirnov statistic yields

$$\|F_n - F\| = O_P(n^{-\frac{1}{2}})$$

so

$$Rem(F_n - F) = o_P(\|F_n - F\|)$$

$$= o_P(n^{-\frac{1}{2}})$$

and condition (2.7) holds.

Huber (1977, 1981) generalized the definition of the Fréchet derivative to include the case where the domain of T is the space M of all probability measures on a finite dimensional Euclidean space and the norm is replaced by a metric that generates the weak topology on M. In this case the metric $d(F_n, F)$ replaces $\|F_n - F\|$ in (2.10) and if $d(F_n, F) = O_P(n^{-\frac{1}{2}})$, then asymptotic normality follows as before.

The use of the Fréchet derivative creates a problem because this derivative is defined on a vector space, and usually statistical functionals are not defined on vector spaces but rather on the space M . Therefore an appropriate extension of a functional to a vector space containing M must be constructed before Fréchet differentiation can be applied to the functional. An advantage of Huber's version of the Fréchet derivative is that it can be applied to a functional without extension to a vector space. Unfortunately this is also a disadvantage, because strong theorems on Fréchet differentiation on vector spaces cannot be applied.

A more serious problem arises from the fact that Fréchet differentiability is such a restrictive condition that frequently statistical functionals are simply not Fréchet differentiable. A classical statistic such as the sample median provides an example.

<u>Example 2.3.2</u>. Consider the vector space $C[0,1]$ of continuous functions on $[0,1]$ with the uniform topology. For $G \in C[0,1]$ define

$$G^{-1}(t) = \inf \{1,x: G(x) \geq t\} \; .$$

For a d.f. with all mass on $[0,1]$, the functional $T(G) = G^{-1}(\frac{1}{2})$ defines a median of G . We shall show that T is not Fréchet differentiable at U , the uniform distribution on $[0,1]$.

The von Mises derivative of T at U is

$$T'_U(G) = -\frac{1}{2} \int_0^{1/2} dG(x) + \frac{1}{2} \int_{1/2}^1 dG(x) \; .$$

If the Fréchet derivative exists, it must coincide with T'_U . Now if we let G_t be the d.f. defined by

$$G_t(x) = \begin{cases} x+t & \text{if } 0 \leq x < \frac{1}{2}-t \\[2ex] \frac{1}{2} & \text{if } \frac{1}{2}-t \leq x \leq \frac{1}{2} \\[2ex] x & \text{if } \frac{1}{2} < x \leq 1 \; , \end{cases}$$

then $T'_U(G_t-U) = 0$. Since $\|G_t-U\| = t$ and $T(G_t) = \frac{1}{2}-t$, we have

$$\frac{T(G_t)-T(U)-T'_U(G_t-U)}{\|G_t-U\|} = -1 \; .$$

so the Fréchet derivative does not exist. Δ

This example suggests that the Fréchet derivative is too strong to be used for statistical functionals. Therefore it is desirable to use a weaker form of derivative for which (2.7) is still valid but which can be applied to large classes of statistics. For this reason Reeds (1976) used the Hadamard derivative, and this is the derivative that we shall consider in these notes.

The norm chosen on the domain of the functional is a crucial factor for the differentiability, and moreover, it is desirable to have a topology which suggests "robustness". The weak topology on the space of probability measures appears to be a natural topology for statistics (according to Hampel (1971)), however stronger topologies such as that induced by the Hellinger metric (which is equivalent to the variation norm, see Kraft (1955)) have been used, and estimators which are differentiable with respect to those topologies have been considered "robust" (see Beran (1977)). The uniform topology, which we shall adopt, is stronger than the weak topology but weaker than the topology induced by the Hellinger metric. With the uniform topology, differentiability will imply "robustness" in the sense of Hampel, as well as asymptotic normality and asymptotic efficiency.

CHAPTER III

HADAMARD DIFFERENTIATION

As we observed in the last chapter, von Mises' method is dependent
upon the differentiability properties of statistical functionals. In
preparation for the study of these properties we present here some basic
definitions and results on differentiation of functions defined on topological vector spaces, and prove an implicit function theorem which will
later be applied to implicitly defined statistical functionals. More
details on the topics covered here can be found in Averbukh and
Smolyanov (1968), Yamamuro (1974), and Keller (1974).

3.1 Definitions of differentiability

It is convenient first to establish a general form of differentiation and then restrict this to the form we wish to use. Let V and W
be topological vector spaces and let $L(V,W)$ be the set of continuous
linear transformations from V to W . Let S be a class of subsets of
V such that every subset consisting of a single point belongs to S ,
and let A be an open subset of V .

Definition 3.1.1. A function $T: A \longrightarrow W$ is S-differentiable at
$F \in A$ if there exists $T_F' \in L(V,W)$ such that for any $K \in S$

$$(3.1) \qquad \lim_{t \to 0} \frac{T(F+tH)-T(F)-T_F'(tH)}{t} = 0$$

16

uniformly for $H \in K$. The linear function T'_F is called the

S-derivative of T at F . Δ

It is convenient to define the remainder term

$$(3.2) \qquad R(T,F,H) = T(F+H)-T(F)-T'_F(H) \ .$$

With this notation (3.1) is equivalent to: for any neighborhood N of 0 in W , there exists $\varepsilon > 0$ such that if $|t| < \varepsilon$ then

$$\frac{R(T,F,tH)}{t} \in N$$

for all $H \in K$.

Here are we interested in three particular types of differentiation:

a) $S = \{$bounded subsets of $V\}$; this corresponds to Fréchet differentiation.

b) $S = \{$compact subsets of $V\}$; this corresponds to Hadamard (or compact) differentiation.

c) $S = \{$single point subsets of $V\}$; this corresponds to Gateaux differentiation.

From these definitions it is clear that Fréchet differentiability implies Hadamard differentiability which in turn implies Gateaux differentiability.

Since the S-derivative defined above must be continuous, a linear function will be S-differentiable if and only if it is continuous. For this reason differentiability is dependent on the topologies of the spaces involved. Moreover, the topology of the domain of a function determines the sets which are in S .

We would like to compare the von Mises derivative with an S-derivative, but we must first overcome two problems. First, the von Mises derivative does not mention continuity, and second, it is not defined on a vector space. If we equip M , the set of probability measures on $\mathbb{R}$,

with the weak topology, that is, the weakest topology for which all
functionals of the form

$$T(F) = \int \phi(x)dF(x) \ , \ F \in M$$

are continuous for ϕ bounded and continuous, then von Mises derivatives
will be continuous on M . Now suppose that we embed M in M^* , the
space of all bounded signed measures on $\mathbb{R}$, where M^* has the weak
topology. If a statistical functional can be extended to M^* , then its
von Mises derivative will correspond to the Gateaux derivative on this
space. Due to this similarity between the two derivatives, the von Mises
derivative has been referred to as the Gateaux derivative in the
statistical literature.

In what follows we shall be particularly interested in Hadamard
differentiation, and we shall adopt a stronger topology than the weak
topology on the domain of our functionals. With a stronger topology
there will be fewer compact sets, so it will be easier to establish
Hadamard differentiability. However, with a stronger topology S-differ-
entiability and von Mises differentiability will not be comparable. But
this is of little importance to us since we have already seen that von
Mises differentiability does not imply the asymptotic results that
interest us.

We shall apply Hadamard and Fréchet differentiation to functions
with domain and range contained in real Banach spaces. These derivatives
are useful because the chain rule holds for them, which is not the case
for the Gateaux derivative.

<u>Proposition 3.1.2.</u> (Chain rule) Let $S = \{$compact subsets$\}$ or
$S = \{$bounded subsets$\}$. Let V , W , Z be topological vector spaces
with $T: V \longrightarrow W$ and $Q: W \longrightarrow Z$. If T is S-differentiable at

$F \in V$ and if Q is S-differentiable at $T(F) \in W$, then $Q \circ T$ is S-differentiable at F and

$$(Q \circ T)'_F = Q'_{T(F)} \circ T'_F \ .$$

Proof: See Yamamuro (1974), p. 11. Note that this proposition also holds when T and Q are defined on appropriate open subsets of V and W , respectively. $\triangle$

3.2 An implicit function theorem

We are interested in establishing the Hadamard differentiability of certain statistical functionals some of which are defined implicitly, so here we shall prove a pointwise implicit function theorem that can be applied in this case. In this section only Hadamard differentiation will be considered, so S = {compact subsets} .

Lemma 3.2.1. Suppose V and W are topological vector spaces with A an open subset of V and let $T: A \longrightarrow W$ be a function. Then T is Hadamard differentiable at $F \in A$ if and only if for every compact set $K \subset V$, every real sequence $\varepsilon_n \longrightarrow 0$, and every sequence $H_n \subset K$,

$$(3.3) \qquad \lim_{n \to \infty} \frac{R(T,F,\varepsilon_n H_n)}{\varepsilon_n} = 0 \ .$$

Proof: Suppose T is Hadamard differentiable at F . Let $K \subset V$ be compact, $\{H_n\} \subset K$, and $\varepsilon_n \longrightarrow 0$. Then for $\delta > 0$

$$N_\delta = \{Y: \|Y\| < \delta\}$$

is a neighborhood of zero in W , so according to the definition of Hadamard differentiation,

$$\|R(T,F,\varepsilon_n H_n)/\varepsilon_n\| < \delta$$

for n large enough. Therefore (3.3) holds.

Conversely, if T is not Hadamard differentiable at F, then there exists a compact set $K \subset V$ and a neighborhood N of zero in W such that for all n, there exists $H_n \in K$ and ε_n with $0 < |\varepsilon_n| < \frac{1}{n}$ such that

$$R(T,F,\varepsilon_n H_n)\varepsilon_n \notin N .$$

Thus

$$\lim_{n \to \infty} R(T,F,\varepsilon_n H_n)/\varepsilon_n \neq 0$$

which contradicts (3.3). $\triangle$

The following definition and theorem will be needed to prove Theorem 3.2.4 below, an implicit function theorem which is applicable to statistical functionals.

<u>Definition 3.2.2</u>. Suppose V and W are topological vector spaces with A an open subset of V and let $T: A \longrightarrow W$ be a function. T is <u>compact preserving</u> at $F \in A$ if for every compact subset $K \subset V$, every real sequence $\varepsilon_n \longrightarrow 0$, and every sequence $\{H_n\} \subset K$, the set

$$\left\{ \frac{T(F+\varepsilon_n H_n)-T(F)}{\varepsilon_n} , \ n \geq 1 \right\}$$

has compact closure in W. $\triangle$

<u>Theorem 3.2.3</u>. (Inverse function theorem). Let A be an open subset of a normed vector space V and let $T: A \longrightarrow V$ be 1-1 and Hadamard differentiable at $F \in A$. Assume that T'_F is 1-1 with continuous inverse. If $S = T^{-1}$ is compact preserving at $G = T(F)$, then S is Hadamard differentiable at G and

$$S'_G = (T'_F)^{-1} .$$

<u>Proof</u>: Let

$$R_1(S,G,X) = S(G+X)-S(G)-(T_F')^{-1}(X) \ .$$

To prove that S is Hadamard differentiable at G with $S_G' = (T_F')^{-1}$,
it suffices to show that for any compact set $K \subset V$, any sequence
$\varepsilon_n \longrightarrow 0$, and any sequence $\{H_n\} \subset K$,

$$\lim_{n\to\infty} R_1(S,G,\varepsilon_n H_n)/\varepsilon_n = 0 \ ,$$

according to the previous lemma. First note that

$$R_1(S,G,X) = -(T_F')^{-1}[R(T,F,S(G+X)-S(G))] \ .$$

Since S is compact preserving at G , there exists a compact set
$K_1 \subset V$ such that

$$\frac{S(G+\varepsilon_n H_n)-S(G)}{\varepsilon_n} \ \epsilon \ K_1$$

for all large n . Since T is Hadamard differentiable at F , Lemma
3.2.1 implies that

$$R(T,F,S(G+\varepsilon_n H_n)-S(G))/\varepsilon_n \longrightarrow 0 \ .$$

Since $(T_F')^{-1}$ is continuous,

$$(T_F')^{-1}[R(T,F,S(G+\varepsilon_n H_n)-S(G))]/\varepsilon_n \longrightarrow 0 \ ,$$

which is equivalent to

$$R_1(S,G,\varepsilon_n H_n)/\varepsilon_n \longrightarrow 0 \ ,$$

so by Lemma 3.2.1 S is Hadamard differentiable at G . $\qquad\qquad \Delta$

<u>Theorem 3.2.4</u> (Implicit function theorem). Let $(G_o,\theta_o) \ \epsilon \ V \times \mathbb{R}^p$, let N
be a neighborhood of G_o and M be a neighborhood of θ_o , and let

$\Psi: N \times M \longrightarrow \mathbb{R}^p$ be Hadamard differentiable at (G_o, θ_o) . Assume that $\Psi(G_o, \theta_o) = 0$, that the partial derivative $D_2\Psi(G_o, \theta_o)$ is non-singular, and that there exists a neighborhood N_o of 0 in $\mathbb{R}^p$ such that $\Psi(G, \theta) = t$ has a unique solution $T(G, t) = \theta$ for $t \in N_o$ and $(G, \theta) \in N \times M$.

If for any compact set $K \subset V \times \mathbb{R}^p$, any real sequence $\varepsilon_n \longrightarrow 0$, and any sequence $\{(H_n, t_n)\} \subset K$,

$$\frac{T(G_o + \varepsilon_n H_n, \varepsilon_n t_n) - T(G_o, 0)}{\varepsilon_n}$$

is bounded for all large n , then the function

$$\tau: N \longrightarrow \mathbb{R}^p$$

defined by $\tau(G) = T(G, 0)$ is Hadamard differentiable at G_o . The derivative of τ is

$$\tau'_{G_o} = -\left(D_2\Psi(G_o, \theta_o)\right)^{-1} \circ D_1\Psi(G_o, \theta_o) \ .$$

<u>Proof</u>: Let $\Lambda: N \times M \longrightarrow V \times \mathbb{R}^p$ be defined by $\Lambda(G, \theta) = (G, \Psi(G, \theta))$. Then Λ has an inverse $\Lambda^{-1}(G, t) = (G, T(G, t))$. Since Ψ is Hadamard differentiable at (G_o, θ_o) it follows that Λ is also Hadamard differentiable at (G_o, θ_o) with derivative

$$\Lambda'(G_o, \theta_o) = \begin{bmatrix} I_V & 0 \\ D_1\Psi(G_o, \theta_o) & D_2\Psi(G_o, \theta_o) \end{bmatrix}$$

where I_V is the identity transformation on V .

Since $D_2\Psi(G_o, \theta_o)$ is non-singular, the linear transformation $\Lambda'(G_o, \theta_o)$ is 1-1 with inverse

$$(3.4) \quad (\Lambda'(G_o,\theta_o))^{-1} = \begin{bmatrix} I_V & 0 \\ -(D_2\Psi(G_o,\theta_o))^{-1}\circ D_1\Psi(G_o,\theta_o) & (D_2\Psi(G_o,\theta_o))^{-1} \end{bmatrix}.$$

This inverse is continuous, so to apply the inverse function theorem, Theorem 3.2.3, we have only to show that Λ^{-1} is compact preserving at $(G_o,0)$.

Let $K \subset V \times \mathbb{R}^p$ be compact, let $\varepsilon_n \longrightarrow 0$ be a real sequence, and let $\{(H_n,t_n)\}$ be a sequence in K . Then

$$\frac{\Lambda^{-1}(G_o+\varepsilon_n H_n,\varepsilon_n t_n)-\Lambda^{-1}(G_o,0)}{\varepsilon_n}$$

$$(3.5) \qquad = (H_n, \frac{T(G_o+\varepsilon_n H_n,\varepsilon_n t_n)-T(G_o,0)}{\varepsilon_n}) .$$

Since $H_n \in \pi_1(K)$, the projection of K into V , which is compact, and since by hypothesis the second component of (3.5) is bounded in $\mathbb{R}^p$, it follows that (3.5) lies in a compact set in $V \times \mathbb{R}^p$. Hence Λ^{-1} is compact preserving at $(G_o,0)$.

The inverse function theorem implies that Λ^{-1} is Hadamard differentiable at $(G_o,0)$ and that

$$(\Lambda^{-1})'(G_o,0) = (\Lambda'(G_o,\theta_o))^{-1} .$$

But

$$(\Lambda^{-1})'(G_o,0) = \begin{bmatrix} I_V & 0 \\ D_1T(G_o,0) & D_2T(G_o,0) \end{bmatrix}$$

so by (3.4) we have

$$\tau'_{G_o} = D_1 T(G_o, 0)$$

$$= -(D_2 \Psi(G_o, \theta_o))^{-1} \circ D_1 \Psi(G_o, \theta_o)$$

and the theorem is proved. $\qquad\qquad\qquad\qquad\qquad\qquad\qquad\qquad \Delta$

CHAPTER IV

SOME PROBABILITY THEORY ON C[0,1] AND D[0,1]

We have seen in Chapter II that to prove asymptotic normality by
von Mises' method it is necessary to show that a statistical functional
is differentiable and that the remainder term of its von Mises expansion
satisfies the convergence condition (2.7). In this chapter we show that
statistical functionals induce functionals on the space D[0,1] of
functions on [0,1] with at most discontinuities of the first kind, and
that problems of differentiability and convergence can be considered in
this setting. Both the differentiability of the functional and the
convergence of the remainder depend on the choice of topology on the
domain of the functional. A stronger topology will allow more functionals to be differentiable, but will interfere with the convergence of the
remainder. We shall use the uniform topology on D[0,1] and we shall
show that with this topology the remainder term satisfies the convergence
condition (2.7). This result will first be proved on C[0,1] , the space
of continuous functions on [0,1] with the uniform topology, and then be
extended to D[0,1] . In the following chapters we shall show that wide
classes of statistical functionals induce Hadamard differentiable
functionals on D[0,1] with the uniform topology, and therefore with
this choice of topology we are able to construct a broadly applicable
von Mises calculus.

The functionals that we shall consider will usually not be defined on the entire space $C[0,1]$ or $D[0,1]$, but rather in a neighborhood of the uniform d.f. U in either of these spaces. Nevertheless, when there is no possibility of confusion, in order to simplify the language we shall refer to such functionals as functionals defined on $C[0,1]$ or $D[0,1]$, respectively, and use the notation $\tau: C[0,1] \longrightarrow \mathbb{R}$ or $\tau: D[0,1] \longrightarrow \mathbb{R}$ to represent them. In fact, this convention will be extended in general to functions between topological vector spaces.

Most of the material in this chapter can be found in Dunford and Schwartz (1958), Billingsley (1968), and Reeds (1976).

4.1 The spaces $C[0,1]$ and $D[0,1]$

Let F be a continuous, strictly increasing d.f. on $\mathbb{R}$. If $X_1,\ldots,X_n$ are i.i.d. random variables with d.f. F , then $F(X_1),\ldots,F(X_n)$ are i.i.d. random variables with d.f. uniform on $[0,1]$. If F_n is the empirical d.f. corresponding to $X_1,\ldots,X_n$ and U_n is the empirical d.f. corresponding to $F(X_1),\ldots,F(X_n)$, it follows that

$$F_n = U_n \circ F \ .$$

If T is a statistical functional, then we can define a functional τ by

$$\tau(U_n) = T(U_n \circ F) = T(F_n)$$

and

$$\tau(U) = T(U \circ F) = T(F) \ .$$

In general for any d.f. G on $[0,1]$, we can define

$$\tau(G) = T(G \circ F)$$

when $T(G \circ F)$ is defined. Therefore for fixed F , the statistical functional T induces a functional τ on the space of d.f.'s with mass concentrated on $[0,1]$. For this reason we can restrict our attention to d.f.'s concentrated on $[0,1]$ and view them as elements of the function spaces $C[0,1]$ and $D[0,1]$, which we shall now consider in detail.

Let $C[0,1]$ be the space of continuous real valued functions on $[0,1]$ with the <u>uniform topology</u>, the topology induced by the sup-norm

$$\|G\| = \sup_{x \in [0,1]} |G(x)| \ , \ G \in C[0,1] \ .$$

Let $D[0,1]$ be the space of right continuous real valued functions on $[0,1]$ which have left hand limits. That is, for $G \in D[0,1]$

$$G(t^+) = \lim_{s \downarrow t} G(s) = G(t) \quad \text{for} \quad 0 \le t < 1 \ ,$$

$$G(t^-) = \lim_{s \uparrow t} G(s) \quad \text{exists for} \quad 0 < t \le 1 \ .$$

A <u>step function</u> $\phi \in D[0,1]$ is a function for which there exists a finite partition $0 = t_0 < t_1 < \ldots < t_n = 1$ such that ϕ is constant on each subinterval $[t_{i-1}, t_i)$ for $i = 1, \ldots, n$.

<u>Lemma 4.1.1</u>. The elements of $D[0,1]$ can be uniformly approximated by step functions.

<u>Proof</u>: Let $G \in D$ and $\epsilon > 0$. We must show that there is a step function ϕ such that $|G(x) - \phi(x)| < \epsilon$ for all $x \in [0,1]$. Such an approximation is possible in some subinterval $[0, t_1)$ because $G(0^+) = G(0)$. Let $a = \sup \{t: G$ can be uniformly approximated by step functions in $[0,t)\}$. Since $G(a^-)$ exists, G can be uniformly approximated on $[0,a]$. If $a < 1$, then, since $G(a) = G(a^+)$, an approximation by step functions can be extended past a , just as was done at 0 . Hence $a = 1$. $\quad\quad\quad \Delta$

This lemma shows that the functions on $D[0,1]$ are bounded, since the step functions are. Therefore we can also define on $D[0,1]$ the norm

$$\|G\| = \sup_{x \in [0,1]} |G(x)|$$

which extends the norm of $C[0,1]$. With the topology induced by this norm, the <u>uniform topology</u>, $D[0,1]$ contains $C[0,1]$ as a closed sub-space.

With the uniform topology, $D[0,1]$ is complete and is therefore a Banach space. However $D[0,1]$ is not separable. The point mass d.f.'s δ_x , $x \in [0,1]$, satisfy $\|\delta_x - \delta_y\| = 1$ for $x \neq y$. Probability theory on $D[0,1]$ is complicated by this fact.

We shall use the Hadamard derivative on the spaces $C[0,1]$ and $D[0,1]$, so we need a characterization of their compact sets.

<u>Proposition 4.1.2.</u> (Arzela-Ascoli). Let $K \subset C[0,1]$. Then K is compact if and only if it is closed, bounded, and equicontinuous.

<u>Proof</u>: This well known theorem is proved, for example, in Dunford and Schwartz (1958), p. 266, Theorem IV.6.7. Recall that K is equicontinuous if for all $\varepsilon > 0$ there is a $\delta > 0$ such that for $s,t \in [0,1]$, if $|s-t| < \delta$ then $|G(s)-G(t)| < \varepsilon$ for all $G \in K$. $\qquad\qquad \Delta$

For $D[0,1]$ the characterization of compact sets is more involved and less well known, so we include a proof adapted from Dunford and Schwartz (1958) (Theorem IV.5.6).

<u>Proposition 4.1.3</u>. A bounded set $K \subset D[0,1]$ has compact closure if and only if for every $\varepsilon > 0$ there exists a partition $0 = t_0 < t_1 < \ldots < t_n = 1$ and points $s_i \in [t_{i-1}, t_i)$, $i = 1,\ldots,n$, such that

$$(4.1) \qquad \sup_{s \in [t_{i-1}, t_i)} |G(s_i) - G(s)| < \varepsilon \ , \ G \in K \ , \ i = 1, \ldots, n \ .$$

<u>Proof</u>: Let A be the set with elements $\alpha = \{t_o, \ldots, t_n, s_1, \ldots, s_n\}$

where $0 = t_o < t_1 < \ldots < t_n = 1$ is a partition of $[0,1]$ and

$s_i \in [t_{i-1}, t_i)$, $i = 1, \ldots, n$. Let A be ordered by the relation

$\alpha' \geq \alpha$ if each interval in α is the union of intervals in α' .

Define $U_\alpha : D[0,1] \longrightarrow D[0,1]$ by

$$U_\alpha(G) = \sum_{i=1}^{n} G(s_i) \, I_{[t_{i-1}, t_i)} \ ,$$

$$U_\alpha(G)(1) = G(1) \ , \qquad\qquad G \in D[0,1] \ ,$$

where I_E stands for the indicator function of a set E . Then

$\|U_\alpha G\| = \|G\|$ and $U_{\alpha'} \circ U_\alpha = U_\alpha$ for $\alpha' \geq \alpha$.

Let $E \subset D[0,1]$ be the family of step functions. Then for $\phi \in E$,
$U_\alpha \phi = \phi$ for α large enough, so $\lim_\alpha U_\alpha \phi = \phi$. Let $\varepsilon > 0$. Since E
is dense in $D[0,1]$ for any $G \in D[0,1]$ there is $\phi \in E$ such that
$\|\phi - G\| < \varepsilon$. Choose α_o large enough such that for $\alpha \geq \alpha_o$, $U_\alpha \phi = \phi$.
Then

$$\|U_\alpha G - G\| \leq \|U_\alpha G - U_\alpha \phi\| + \|\phi - G\| < 2\varepsilon \ ,$$

so $\lim_\alpha U_\alpha G = G$ for all $G \in D[0,1]$.

Now suppose that $K \subset D[0,1]$ has compact closure. Then for $\varepsilon > 0$
there are $G_1, \ldots, G_n \in K$ such that

$$\inf_i \|G_i - G\| < \varepsilon \ \text{ for all } \ G \in K \ .$$

Choose α_o large enough such that for $\alpha \geq \alpha_o$,

$$\|U_\alpha G_i - G_i\| < \varepsilon \ , \ i = 1, \ldots, n \ .$$

Then for $G \in K$ there is G_i such that

$$\|U_\alpha G - G\| \leq \|U_\alpha G - U_\alpha G_i\| + \|U_\alpha G_i - G_i\| + \|G_i - G\|$$

$$< 3\epsilon .$$

Therefore the partition associated with α satisfies (4.1).

Conversely, suppose K is bounded and for $\epsilon > 0$ we can find $\alpha_0 \in A$ such that if $\alpha \geq \alpha_0$, then (4.1) holds. Then for $\alpha \geq \alpha_0$, $\|U_\alpha G - G\| < \epsilon$ for all $G \in K$. Now $U_\alpha K$ is finite dimensional and bounded, and hence has compact closure, so we can find $G_1, \ldots, G_n \in K$ such that

$$\inf_i \|U_\alpha G_i - U_\alpha G\| < \epsilon \text{ for all } G \in K .$$

Therefore for $G \in K$, there is a G_i such that

$$\|G - G_i\| < \|G - U_\alpha G\| + \|U_\alpha G - U_\alpha G_i\| + \|U_\alpha G_i - G_i\|$$

$$< 3\epsilon .$$

Hence K has compact closure. $\triangle$

Since $C[0,1]$ is a subspace of $D[0,1]$, the compact subsets of $C[0,1]$ are also compact in $D[0,1]$.

4.2 Probability theory on $C[0,1]$

Since $C[0,1]$ is a separable Banach space, probability theory on it is somewhat simpler than on $D[0,1]$. For this reason we shall consider it first.

Let V be a topological vector space with Borel σ-field $\mathcal{B}$ generated by the open sets of V. Let $(\Omega, \mathcal{F}, P)$ be a probability space with probability measure P defined on the σ-field $\mathcal{F}$ of measurable subsets of Ω. A <u>random element</u> of V is a measurable function

$$Y: \Omega \longrightarrow V \; ,$$

measurable in the sense that $Y^{-1}\mathcal{B} \subset F$. The random element Y defines a probability measure P_o on V by the relation

$$P_o(B) = P\{\omega \in \Omega : Y(\omega) \in B\}$$

for $B \in \mathcal{B}$. A sequence $\{Y_n ; n \geq 1\}$ of random elements (or measurable functions) on V <u>converges in distribution</u> to Y , and we write

$$Y_n \xrightarrow{\;\mathcal{D}\;} Y \; ,$$

if the corresponding probability measures P_n on V converge weakly to P_o , the measure corresponding to Y .$'$ P_n <u>converges weakly</u> to P_o if for every bounded continuous real valued function f on V ,

$$\lim_{n \to \infty} \int_V f \; dP_n = \int_V f \; dP_o \; .$$

If $\mathcal{P}$ is a family of probability measures on V then $\mathcal{P}$ is <u>tight</u> if for any $\varepsilon > 0$ there is a compact set $K \subset V$ such that $P(K) > 1-\varepsilon$ for all $P \in \mathcal{P}$. $\mathcal{P}$ is <u>relatively compact</u> if every sequence of elements in $\mathcal{P}$ contains a weakly convergent subsequence. The following theorem, due to Prohorov, is of central importance.

<u>Theorem 4.2.1</u>. (i) If $\mathcal{P}$ is tight, then it is relatively compact.

(ii) If V is separable and complete and if $\mathcal{P}$ is relatively compact, then $\mathcal{P}$ is tight.

<u>Proof</u>: See Billingsley (1968) p. 37. $\qquad\qquad\qquad\qquad\qquad\qquad \Delta$

Now suppose that (Ω, F, P) is a probability space and let $Y_1, \ldots, Y_n$ be i.i.d. random variables defined on Ω and having the uniform distribution on $[0,1]$. Let U_n be the empirical d.f. on $[0,1]$ corresponding to $Y_1, \ldots, Y_n$. We wish to study U_n in the setting of $C[0,1]$, and since U_n is not continuous we must modify it somewhat. Following

Billingsley (1968), let U_n^* be the continuous version of U_n where U_n^*

is the d.f. corresponding to a uniform distribution of mass $(n+1)^{-1}$ in

each of the $n+1$ intervals $[Y_{(i-1)},Y_{(i)}]$ where $Y_{(1)} \leq Y_{(2)} \leq \cdots \leq Y_{(n)}$

are the ordered Y_i's , $Y_{(0)} = 0$, and $Y_{(n+1)} = 1$. With probability

one, $Y_i \neq Y_j$ for $i \neq j$, so U_n^* is continuous and

$$\|U_n^* - U_n\| \leq \frac{1}{n} \qquad \text{a.s.} .$$

Since $C[0,1]$ is separable, it can be shown that a function

$Y: \Omega \longrightarrow C[0,1]$ is measurable if and only if its cross-sections

$Y(t) = Y(\omega,t)$ are measurable. From this we see that U_n^* is a random

element of $C[0,1]$.

Consider now the random elements $Z_n = \sqrt{n}(U_n^* - U)$ of $C[0,1]$ where

U is the uniform d.f. $U(x) = x$. Work due to Doob (1949) and Donsker

(1952) shows that as $n \longrightarrow \infty$,

$$Z_n \xrightarrow{\;\mathcal{D}\;} W^o \;,$$

where W^o is the Brownian bridge, the Gaussian stochastic process

defined on $[0,1]$ satisfying

$$EW^o(t) = 0 \;,\; t \in [0,1] \;,$$

$$EW^o(s)W^o(t) = s(1-t) \;,\; s \leq t \;,\; s,t, \in [0,1] \;.$$

W^o is also a random element of $C[0,1]$. For discussion and proofs of

this material, see Billingsley (1968).

We wish to consider $\{P_n, n \geq 1\}$, the sequence of probability

measures corresponding to Z_n , $n \geq 1$. Since $Z_n \xrightarrow{\;\mathcal{D}\;} W^o$, the sequence

P_n converges weakly to P_o , the probability measure corresponding to

W^o . It follows that the family $\mathcal{P} = \{P_o, P_n, n \geq 1\}$ is relatively compact, and since $C[0,1]$ is separable and complete, P is tight by Prohorov's theorem.

The fact that the family P is tight can be combined with Hadamard differentiability to prove that the remainder term of the von Mises expansion of a statistical functional satisfies the convergence condition (2.7). This approach was developed by Reeds (1976) and we shall follow it here.

Suppose that a functional $\tau: C[0,1] \longrightarrow \mathbb{R}$ is Hadamard differentiable at U . According to Definition 3.1.1, if we define the function Rem by

$$(4.2) \qquad \mathrm{Rem}(tH) = \tau(U+tH) - \tau(U) - \tau'_U(tH) ,$$

then for any compact $K \subset C[0,1]$,

$$(4.3) \qquad \lim_{t \to 0} \frac{\mathrm{Rem}(tH)}{t} = 0$$

uniformly for $H \in K$.

Proposition 4.2.2. If for any compact set $K \subset C[0,1]$,

$$\lim_{t \to 0} \frac{\mathrm{Rem}(tH)}{t} = 0$$

uniformly for $H \in K$, then

$$\sqrt{n} \; \mathrm{Rem}(U_n^* - U) \xrightarrow{\;P\;} 0 .$$

Proof: Let $\epsilon > 0$. Let P_n be the probability measure on $C[0,1]$ corresponding to $\sqrt{n}(U_n^* - U)$ such that for $A \subset C[0,1]$,

$$P_n(A) = P\{\sqrt{n}(U_n^* - U) \in A \} .$$

Then Prohorov's theorem implies that $\{P_n , n > 1\}$ is tight, so there exists a compact subset $K \subset C[0,1]$ such that for all n ,

$$P_n(K) > 1-\varepsilon \quad .$$

It follows from (4.3) that there exists an n_o such that for $n > n_o$ and $H \in K$,

$$\left| \sqrt{n}\ \mathrm{Rem}(\tfrac{1}{\sqrt{n}} H) \right| < \varepsilon \quad .$$

Therefore

$$P\{\left| \sqrt{n}\ \mathrm{Rem}(\tfrac{1}{\sqrt{n}}[\sqrt{n}(U_n^*-U)]) \right| < \varepsilon\} > 1-\varepsilon$$

for all $n > n_o$. Since $\varepsilon > 0$ was arbitrary,

$$\sqrt{n}\ \mathrm{Rem}(U_n^*-U) \xrightarrow{P} 0 \quad . \qquad\qquad \Delta$$

4.3 Probability theory on $D[0,1]$

Since the empirical d.f.'s are elements of $D[0,1]$ but not of $C[0,1]$, most of the functionals that we shall study will be defined on $D[0,1]$ rather than $C[0,1]$. However, since $D[0,1]$ is not separable, complications arise. One such complication is the fact that the empirical d.f.'s U_n corresponding to a sample of uniformly distributed random variables on $[0,1]$ are not random elements of $D[0,1]$. To see this, consider the case $n = 1$. Let (Ω,F,P) be a probability space and let $Y_1 : \Omega \longrightarrow [0,1]$ be a random variable uniformly distributed on $[0,1]$. If U_1 is the corresponding empirical d.f., then

$$(4.4) \qquad\qquad U_1(\omega) = \delta_{Y_1(\omega)} \quad ,$$

the d.f. of the point mass at $Y_1(\omega)$. We shall show that the function

$$U_1 : \Omega \longrightarrow D[0,1]$$

defined by (4.4) is not measurable.

The random variable Y_1 induces a probability measure μ on $[0,1]$ by $\mu(B) = P\{Y_1(\omega) \in B\}$ for any Borel set $B \subset [0,1]$. Since Y_1 is uniformly distributed, μ coincides with the Lebesgue measure on $[0,1]$. Now for $x \in [0,1]$, let $N_x = \{G \in D[0,1]: \|G - \delta_x\| < \frac{1}{2}\}$. N_x is open in $D[0,1]$, so for any subset $B \subset [0,1]$,

$$O_B = \bigcup_{x \in B} N_x$$

is also open. For any $x \in [0,1]$, $Y_1(\omega) = x$ if and only if $\delta_{Y_1(\omega)} \in N_x$, so if U_x is a measurable function then

$$P\{Y_1(\omega) \in B\} = P\{U_1(\omega) \in O_B\}$$

for any set $B \subset [0,1]$. But then all subsets of $[0,1]$ are Lebesgue measurable, which is false.

Since the empirical d.f.'s are not random elements of $D[0,1]$, it is convenient to study them by means of the modified d.f.'s U_n^* in $C[0,1]$ that we considered in the preceding section, an approach due to Reeds (1976).

Let $H \in D[0,1]$ and $K \subset D[0,1]$ and define

$$\text{dist}(H,K) = \inf_{G \in K} \|H - G\| .$$

Lemma 4.3.1. Let $Q: D[0,1] \times \mathbb{R} \longrightarrow \mathbb{R}$ and suppose that for any compact set $K \subset D[0,1]$

$$\lim_{t \to 0} Q(H,t) = 0$$

uniformly for $H \in K$. Let $\epsilon > 0$, and let δ_n be a sequence of numbers such that $\delta_n \downarrow 0$. Then for any compact set $K \subset D[0,1]$ there exists n_0 such that for $n > n_0$, if $\mathrm{dist}(H,K) \leq \delta_n$ then

$$|Q(H,\delta_n| < \epsilon .$$

<u>Proof</u>: Suppose not. Then for $\epsilon > 0$, there exists $K \subset D[0,1]$ compact, $\delta_n \downarrow 0$, and a sequence $\{H_n\} \subset D[0,1]$ with $\mathrm{dist}(H_n,K) \leq \delta_n$ such that

$$|Q(H_n,\delta_n| \geq \epsilon$$

for infinitely many n.

Choose a subsequence $\{H_{n_i}\}$ of $\{H_n\}$ such that

$$(4.5) \qquad |Q(H_{n_i},\delta_{n_i})| \geq \epsilon \quad \text{for all } i .$$

Now $\mathrm{dist}(H_{n_i},K) \leq \delta_{n_i}$, so we can choose $H_{n_i}^* \in K$ such that $\|H_{n_i} - H_{n_i}^*\| \leq \delta_{n_i}$. Then $\{H_{n_i}^*\}$ has an accumulation point $H^* \in K$. Therefore we can choose a subsequence of $\{H_{n_i}^*\}$, also denoted by $H_{n_i}^*$, such that $H_{n_i}^* \longrightarrow H^*$. But then $H_{n_i} \longrightarrow H^*$ also, and the set

$$K_1 = \bigcup_{i=1}^{\infty} \{H_{n_i}\} \cup \{H^*\}$$

is compact. Hence $Q(H_{n_i},t) \longrightarrow 0$ uniformly for $H_{n_i} \in K_1$ as $t \longrightarrow 0$. This contradicts (4.5). $\qquad\qquad \Delta$

Since the functions $\sqrt{n}(U_n - U)$ are not random elements of $D[0,1]$ we shall use the inner probability P_* corresponding to P to deal with them.

<u>Lemma 4.3.2.</u> For $\varepsilon > 0$ there exists a compact set $K \subset D[0,1]$ such that for all n ,

$$P_*\{\mathrm{dist}(\sqrt{n}(U_n-U),K) \leq \tfrac{1}{\sqrt{n}}\} > 1-\varepsilon .$$

<u>Proof</u>: Recall that $\|U_n^*-U_n\| < \tfrac{1}{n}$. As in Proposition 4.2.2, there exists a compact set $K \subset C[0,1]$ such that for all n ,

$$P\{\sqrt{n}(U_n^*-U) \in K \} > 1-\varepsilon .$$

Since $C[0,1] \subset D[0,1]$, K is also compact in $D[0,1]$. If $\sqrt{n}(U_n^*-U) \in K$ and $\|U_n^*-U_n\| < \tfrac{1}{n}$, then

$$\mathrm{dist}(\sqrt{n}(U_n-U),K) \leq \tfrac{1}{\sqrt{n}} ,$$

so

$$P_*\{\mathrm{dist}(\sqrt{n}(U_n-U),K) \leq \tfrac{1}{\sqrt{n}}\} > 1-\varepsilon . \qquad \Delta$$

Suppose that T is a statistical functional with induced functional $\tau: D[0,1] \longrightarrow \mathbb{R}$. If τ is Hadamard differentiable at the uniform d.f. U , then the remainder term defined in equation (4.2) satisfies

$$\mathrm{Rem}(U_n-U) = \tau(U_n) - \tau(U) - \tau'_U(U_n-U) .$$

We shall show in the next section that although (U_n-U) may not be a measurable element of $D[0,1]$, nevertheless $\mathrm{Rem}(U_n-U)$ is a measurable function, and therefore we can use the probability measure P for events concerning this function.

<u>Proposition 4.3.3.</u> If for any compact set $K \subset D[0,1]$

$$\frac{\mathrm{Rem}(tH)}{t} \longrightarrow 0 \quad \text{as} \quad t \longrightarrow 0$$

uniformly for $H \in K$, then

$$\sqrt{n} \; \mathrm{Rem}(U_n - U) \xrightarrow{P} 0 \;\;.$$

<u>Proof</u>: Let $\varepsilon > 0$. By Lemma 4.3.2 we can choose a compact $K \subset D[0,1]$ such that

$$P_*\{\mathrm{dist}(\sqrt{n}(U_n - U), K) \leq \frac{1}{\sqrt{n}}\} > 1 - \varepsilon/2 \;\;.$$

Therefore we can find measurable sets E_n , for all n , such that

$$E_n \subset \{\mathrm{dist}(\sqrt{n}(U_n - U), K) \leq \frac{1}{\sqrt{n}}\}$$

and

$$P\{E_n\} > 1 - \varepsilon \;\;.$$

If we apply Lemma 4.3.1 with $Q(H,t) = \dfrac{\mathrm{Rem}(tH)}{t}$, then there exists n_o such that for $n > n_o$, $\mathrm{dist}(H,K) \leq \dfrac{1}{\sqrt{n}}$ implies that

$$\left| \sqrt{n} \; \mathrm{Rem}(\frac{1}{\sqrt{n}} H) \right| < \varepsilon \;\;.$$

Therefore for $n > n_o$ and $H = \sqrt{n}(U_n - U)$,

$$P\{\left| \sqrt{n} \; \mathrm{Rem}(U_n - U) \right| < \varepsilon\} \geq P\{E_n\} > 1 - \varepsilon \;\;.$$

Hence

$$\sqrt{n} \; \mathrm{Rem}(U_n - U) \xrightarrow{P} 0 \;\;. \qquad\qquad \Delta$$

4.4 Asymptotic Normality

In Chapter II it was shown that neither the existence of the influence curve nor that of the von Mises derivative was enough to imply the asymptotic normality of a statistical functional. Here we shall show that a statistical functional is asymptotically normal if the functional it induces on $D[0,1]$ has nonzero Hadamard derivative.

Let $X_1,\ldots,X_n$ be i.i.d. random variables with d.f. F which we assume to be continuous and strictly increasing. As we saw in section 4.1, T induces $\tau: D[0,1] \longrightarrow \mathbb{R}$ by

$$\tau(G) = T(G \circ F)$$

for all $G \in D[0,1]$ such that $T(G \circ F)$ is defined. In particular if $U_n = F_n \circ F^{-1}$ and U is the uniform distribution on $[0,1]$, we have

$$\tau(U_n) = T(F_n)$$

and

$$\tau(U) = T(F) .$$

<u>Lemma 4.4.1</u>. Let T and τ be as above and suppose that τ has Gateaux derivative τ'_U at U. Then T has influence curve

$$IC(x;F,T) = \tau'_U((\delta_x - F) \circ F^{-1}) .$$

<u>Proof</u>: For any d.f. H, $T(H) = \tau(H \circ F^{-1})$ so the influence curve for T at F satisfies

$$IC(x;F,T) = \lim_{t \to 0} \frac{T(F + t(\delta_x - F)) - T(F)}{t}$$

$$= \lim_{t \to 0} \frac{\tau(U + t(\delta_x - F) \circ F^{-1}) - \tau(U)}{t}$$

$$= \tau'_U((\delta_x - F) \circ F^{-1})$$

since τ has Gateaux derivative τ'_U at U. $\qquad\qquad \Delta$

Since Hadamard differentiability is stronger than Gateaux differentiability, the existence of the Hadamard derivative also implies the existence of the influence curve.

<u>Theorem 4.4.2.</u> Let T be a statistical functional and suppose that X is a random variable with continuous, strictly increasing d.f. F . Let τ be the functional induced on $D[0,1]$ by $\tau(G) = T(G \circ F)$, $G \in D[0,1]$. If τ is Hadamard differentiable at U and if

$$0 < \sigma^2 = \mathrm{Var}_F \, IC(X;F,T) < \infty \; ,$$

then

$$\sqrt{n}(T(F_n) - T(F)) \xrightarrow{\;\mathcal{D}\;} N(0,\sigma^2) \; .$$

<u>Proof</u>: Consider

$$\sqrt{n}(T(F_n) - T(F)) = \sqrt{n}(\tau(U_n) - \tau(U))$$

$$(4.6) \qquad\qquad = \sqrt{n}\, \tau'_U(U_n - U) + \sqrt{n}\, \mathrm{Rem}(U_n - U) \; .$$

Now,

$$\tau'_U(U_n - U) = \tau'_U((F_n - F) \circ F^{-1})$$

$$= \tau'_U\!\left(\frac{1}{n} \sum_{1}^{n} (\delta_{X_i} - F) \circ F^{-1}\right)$$

$$(4.7) \qquad\qquad = \frac{1}{n} \sum_{1}^{n} IC(X_i;F,T)$$

by Lemma 4.4.1, so $\tau'_U(U_n - U)$ is a random element of $D[0,1]$. Since $\tau(U_n) = T(F_n)$ is also measurable, we can conclude from (4.6) that $\mathrm{Rem}(U_n - U)$ is a random element of $D[0,1]$. By Proposition 4.3.3, we have

$$\sqrt{n}\, \mathrm{Rem}(U_n - U) \xrightarrow{\;P\;} 0 \; .$$

By combining (4.6) and (4.7) we obtain

$$\sqrt{n}(T(F_n) - T(F)) = \frac{1}{\sqrt{n}} \sum_{1}^{n} IC(X_i;F,T) + \sqrt{n}\, \mathrm{Rem}(U_n - U) \; ,$$

and the theorem is proved by applying the central limit theorem and Slutsky's lemma. Δ

Suppose now that a statistical functional T induces a functional τ on $C[0,1]$ defined by $\tau(G) = T(G \circ F)$, $G \in C[0,1]$, for some fixed d.f. F , and that $\tau(U_n^*) = T(F_n)$ where U_n^* is the continuous version of $U_n = F_n \circ F^{-1}$ defined in section 4.2. In this case we have

<u>Corollary 4.4.3</u>. Let T be a statistical functional and suppose that X is a random variable with continuous, strictly increasing d.f. F . Let τ be the functional induced by T on $C[0,1]$ satisfying $\tau(U) = T(F)$ and $\tau(U_n^*) = T(F_n)$. If τ is Hadamard differentiable at U and if T has influence curve $IC(x;F,T)$ with

$$0 < \sigma^2 = \text{Var}_F \, IC(X;F,T) < \infty \quad ,$$

then

$$\sqrt{n}(T(F_n) - T(F)) \xrightarrow{\;\mathcal{D}\;} N(0,\sigma^2) \ .$$

<u>Proof</u>: We have

$$\sqrt{n}(T(F_n) - T(F)) = \sqrt{n} \ (\tau(U_n^*) - \tau(U))$$

$$= \sqrt{n} \ \tau'_U(U_n^* - U) + \sqrt{n} \ \text{Rem}(U_n^* - U) \ .$$

By Proposition 4.2.2, $\sqrt{n} \ \text{Rem}(U_n^* - U) \xrightarrow{\;P\;} 0$, so it remains to show that

$$\sqrt{n} \ \tau'_U(U_n^* - U) \xrightarrow{\;\mathcal{D}\;} N(0,\sigma^2) \ .$$

Since τ'_U is linear and continuous on $C[0,1]$, there exists a regular Borel measure m on $[0,1]$ such that

$$(4.8) \qquad\qquad \tau'_U(G) = \int_0^1 G(x) \, dm(x)$$

for all $G \in C[0,1]$ (by the Riesz representation theorem, see Dunford and Schwartz (1958) p. 265 Theorem 3). This measure defines a continuous linear functional on $D[0,1]$ which we shall also denote by τ_U' , and which is defined by (4.8) for all $G \in D[0,1]$. Since $\|U_n^* - U_n\| \le 1/n$ a.s.,

$$\sqrt{n}(U_n^* - U_n) \longrightarrow 0 \qquad \text{a.s.}$$

as $n \longrightarrow \infty$, so by continuity of τ_U' ,

$$\sqrt{n}\, \tau_U'(U_n^* - U_n) \longrightarrow 0 \qquad \text{a.s.} \ .$$

Thus we need only to prove that

$$\sqrt{n}\, \tau_U'(U_n - U) \xrightarrow{\ \mathcal{D}\ } N(0,\sigma^2) \ ,$$

which follows as in Theorem 4.4.2. $\qquad\qquad\qquad\qquad\qquad\qquad\qquad\Delta$

CHAPTER V

M-, L-, AND R-ESTIMATORS

In this chapter we shall introduce the three basic types of robust estimators, M-, L-, and R-estimators, and shall study properties of the corresponding statistical functionals. The results established here will later be used to show that the functionals induced on $D[0,1]$ by these estimators are Hadamard differentiable. As we have seen in the previous chapter, the Hadamard differentiability of the induced functionals on $D[0,1]$ is sufficient to imply the asymptotic normality of the estimators.

The asymptotic distribution of these three types of estimators has been studied by various authors using different variations of von Mises' method. M-, and L-estimators were considered by Reeds (1976), Boos (1979), and Boos and Serfling (1980), with different types of derivatives used in each case. R-estimators were considered by Fernholz (1979) following the approach presented here.

5.1 M-estimators

Given a function $\rho(x,\theta)$ and a sample $X_1,\ldots,X_n$, an estimator $T_n = T_n(X_1,\ldots,X_n)$ which minimizes an equation of the form

$$(5.1) \qquad \sum_{i=1}^{n} \rho(X_i,T_n)$$

is called an M-estimator, or maximum likelihood type estimator. When

$$\psi(x,\theta) = \frac{\partial}{\partial\theta}\, \rho(x,\theta)\ ,$$

equation (5.1) can be replaced by

$$(5.2) \qquad\qquad \sum_{i=1}^{n} \psi(X_i, T_n) = 0\ ,$$

and the M-estimator is defined implicitly as a solution of this equation.
In particular, when

$$\rho(x,\theta) = \log\, f(x,\theta)$$

for the population density f , then T_n is the usual maximum likelihood estimator of θ .

This class of statistics was first considered by Huber (1964) and
is treated in greater detail in Huber (1981). We shall be interested in
M-estimators of location which correspond to

$$\psi(x,\theta) = \psi(x-\theta)\ ,$$

so we shall consider estimators of the form $T_n = \theta$ where θ is a
solution of

$$(5.3) \qquad\qquad \sum_{i=1}^{n} \psi(X_i-\theta) = 0\ .$$

The functional corresponding to (5.3) is defined to be a root $T(F) = \theta$
of

$$\int \psi(x-\theta)\,dF(x) = 0\ ,$$

or equivalently, of

$$(5.4) \qquad\qquad \int \psi(F^{-1}(x)-\theta)\,dx = 0$$

where the population of d.f. F is strictly increasing and continuous.
For our purposes we shall define an <u>M-estimator</u> to be a root $T(F_n) = \theta$
of

$$(5.5) \qquad \int \psi(F_n^{-1}(x)-\theta)dx = 0 \ .$$

The functional T defined implicitly by (5.4) induces a functional
τ on $D[0,1]$ by the relation $\tau(G) = T(G{\circ}F)$ for $G \in D[0,1]$, as was
shown in Chapter IV. Hence $\tau(G) = \theta$ is a root of

$$\int_0^1 \psi(F^{-1}(G^{-1}(x))-\theta)dx = 0 \ .$$

In order to study the differentiability of τ , it is convenient to
introduce the function $\Phi: D[0,1] \times \mathbb{R} \longrightarrow \mathbb{R}$ defined by

$$\Phi(G,\theta) = \int_0^1 \psi(F^{-1}(G^{-1}(x))-\theta)dx \ .$$

In Chapter VII we shall prove that Φ is Hadamard differentiable at
(U,θ_0) where U is the uniform d.f. on $[0,1]$ and θ_0 satisfies
$\Phi(U,\theta_0) = 0$, and then apply an implicit function theorem to show that
τ itself is Hadamard differentiable.

The result that follows, Theorem 5.1.2, is a standard form of
theorem which permits the application of Theorem 6.2.1, the implicit
function theorem for functionals on $D[0,1]$ or $C[0,1]$. The conditions
established for M-estimators in Theorem 5.1.2, conditions i), ii), and
iii), are precisely the conditions needed in Theorem 6.2.1. These three
conditions will later be established for other implicitly defined
statistical functionals that we shall consider.

In what follows, we shall use the mean value theorem in the
following form:

<u>Lemma 5.1.1</u>. Let $g: \mathbb{R} \longrightarrow \mathbb{R}$ be continuously differentiable and let $a < b$. Then there exists a Borel measurable function $\beta(x)$ defined on $\mathbb{R}$ such that $a \leq \beta(x) \leq b$ for all $x \in \mathbb{R}$ and

$$g(b+x)-g(a+x) = g'(\beta(x)+x)(b-a)$$

for all $x \in \mathbb{R}$.

<u>Proof</u>: The function $\lambda: \mathbb{R}^2 \longrightarrow \mathbb{R}$ defined by

$$\lambda(t,x) = g'(t+x) - \frac{g(b+x)-g(a+x)}{b-a}$$

is continuous, so $\lambda^{-1}(0) \subset \mathbb{R}^2$ is a closed set. By the mean value theorem,

$$\lambda^{-1}(0) \cap ([a,b] \times \{x\}) \neq \emptyset , \quad \text{for all} \quad x \in \mathbb{R} .$$

Let

$$\beta(x) = \inf \{t \in [a,b]: (t,x) \in \lambda^{-1}(0)\} .$$

Since $\lambda^{-1}(0) \cap ([a,b] \times \mathbb{R})$ is closed, if $\beta(x_o) > y_o$ then $\beta(x) > y_o$ in some neighborhood of x_o . Therefore β is lower semi-continuous and hence Borel measurable. Δ

<u>Theorem 5.1.2</u>. Let F be a continuous d.f. with piecewise continuous density $f = F' > 0$. Let ψ be bounded, continuous, nondecreasing, and piecewise differentiable, with bounded derivative ψ' such that $\psi'(x) \geq m > 0$, m constant, for x in some neighborhood of 0 . Then

    i) for $G,H \in D[0,1]$, $G \geq H$ implies that

$$\Phi(G,\theta) \leq \Phi(H,\theta)$$

for all $\theta \in \mathbb{R}$;

ii) if U is the uniform d.f. and if $\theta_O \in \mathbb{R}$ satisfies

$\Phi(U,\theta_O) = 0$, then there are neighborhoods N_O of θ_O and N_U of U and positive constants A and B such that

$$A(\theta-\sigma) \le \Phi(G,\sigma)-\Phi(G,\theta) \le B(\theta-\sigma)$$

for all $\theta,\sigma \in N_O$ with $\theta \ge \sigma$ and $G \in N_U$:

iii) there exists a constant $k > 0$ such that for $G,H \in D[0,1]$,

$$|\Phi(G,\theta)-\Phi(H,\theta)| \le k \, \|G-H\|$$

for all $\theta \in \mathbb{R}$.

<u>Proof</u>: i) If $G \ge H$ then $G^{-1} \le H^{-1}$. Since F^{-1} and ψ are non-decreasing, we have

$$\psi(F^{-1}(H^{-1}(x))-\theta) \ge \psi(F^{-1}(G^{-1}(x))-\theta) ,$$

so it follows that $\Phi(H,\theta) \ge \Phi(G,\theta)$.

ii) Let $\theta \ge \sigma$. Then by Lemma 5.1.1

$$(5.6) \qquad \Phi(G,\sigma)-\Phi(G,\theta) = (\theta-\sigma)\int_0^1 \psi'[F^{-1}(G^{-1}(x))-\beta(F^{-1}(G^{-1}(x)))]dx$$

where β is Borel measurable with $\sigma \le \beta(x) \le \theta$. Since ψ' is bounded, say by $B > 0$, we have

$$\Phi(G,\sigma)-\Phi(G,\theta) \le B(\theta-\sigma) .$$

Now let $\varepsilon > 0$ and suppose that for $|x| < 2\varepsilon$, $\psi'(x) \ge m$ for some $m > 0$. Let $x_O = F(\theta_O)$. If $N_O = \{\theta: |\theta-\theta_O| < \varepsilon\}$ then for $\sigma,\theta \in N_O$, we have

$$\beta(F^{-1}(G^{-1}(x)) \in N_O ,$$

where β is as in equation (5.6).

48

Since F is strictly increasing and continuous, F^{-1} is continuous, so there is $\delta > 0$ such that if $|x - x_0| < 2\delta$ then $|F^{-1}(x) - \theta_0| < \varepsilon$. We can assume that $\delta \leq \varepsilon$. Let $N_U = \{G: \|G - U\| < \delta\}$. If $\|G - U\| < \delta$ then $\|G^{-1} - U\| < \delta$, so for $G \in N_U$ and $|x - x_0| < \delta$,

$$|G^{-1}(x) - x_0| \leq |G^{-1}(x) - x| + |x - x_0|$$

$$\leq 2\delta .$$

Hence if $|x - x_0| < \delta$ then $|F^{-1}(G^{-1}(x)) - \theta_0| < \varepsilon$, so

$$|F^{-1}(G^{-1}(x)) - \beta(F^{-1}(G^{-1}(x)))|$$

$$\leq |F^{-1}(G^{-1}(x)) - \theta_0| + |\theta_0 - \beta(F^{-1}(G^{-1}(x)))|$$

$$< 2\varepsilon .$$

Therefore, for $|x - x_0| < \delta$,

$$\psi'[F^{-1}(G^{-1}(x)) - \beta(F^{-1}(G^{-1}(x)))] \geq m ,$$

thus

$$\Phi(G,\sigma) - \Phi(G,\theta) \geq A(\theta - \sigma)$$

for $A = \delta m$.

iii) Let $\varepsilon > 0$ and suppose that $G, H \in D[0,1]$ with $\|G - H\| < \varepsilon$. For a given $\theta \in \mathbb{R}$, we can assume without loss of generality that $\Phi(G,\theta) \leq \Phi(H,\theta)$. Since $H \geq G - \varepsilon$, we have

$$|\Phi(G,\theta) - \Phi(H,\theta)| = \Phi(H,\theta) - \Phi(G,\theta)$$

$$\leq \Phi(G-\varepsilon,\theta) - \Phi(G,\theta)$$

$$= \int_0^1 \psi[F^{-1}(G^{-1}(x+\epsilon))-\theta]dx - \Phi(G,\theta)$$

$$= \int_\epsilon^{1+\epsilon} \psi[F^{-1}(G^{-1}(x))-\theta]dx - \Phi(G,\theta)$$

$$\leq 2\epsilon \sup |\psi| \ .$$

Therefore iii) holds with $k = 2 \sup |\psi|$. Δ

The M-estimator $T(F_n)$ defined by (5.5) is not scale invariant. To obtain a scale invariant version of this estimator, we can replce equation (5.5) by

$$(5.7) \qquad \int_0^1 \psi[\frac{F_n^{-1}(x)-\theta}{S_n}]dx = 0$$

where S_n is an estimator of scale. Corresponding to this estimator, we can consider $\Phi: D[0,1] \times \mathbb{R} \longrightarrow \mathbb{R}$ defined by

$$\Phi(G,\theta) = \int_0^1 \psi[\frac{F^{-1}(G^{-1}(x))-\theta}{S(G \circ F)}] \ dx \ .$$

If $S(G \circ F)$, with fixed F , is non-vanishing and Hadamard differentiable at $G = U$, then this type of estimator can be treated in the same manner as the simple M-estimator above.

5.2 L-estimators

A linear combination of a function of order statistics is called an L-estimator. If $X_1,\ldots,X_n$ is a random sample from a d.f. F , then an L-estimator is a statistic of the form

$$(5.8) \qquad T_n = \sum_{i=1}^n w_{ni} h(X_{(i)}) \ ,$$

where $X_{(i)}$ is the i-th order statistic of $X_1,\ldots,X_n$, h is a real valued function, and the weights w_{ni} are real numbers. If the weights are generated by a measure $dM(x) = m(x)dx$ on $[0,1]$ by

$$w_{ni} = \int_{\frac{i-1}{n}}^{\frac{i}{n}} m(x)dx \ ,$$

then (5.8) can be written

$$(5.9) \qquad T_n = \int_0^1 h(F_n^{-1}(x))m(x)dx$$

where F_n is the empirical d.f. for $X_1,\ldots,X_n$. For our purposes, we define an __L-estimator__ to be an estimator of the form (5.9). Such an L-estimator is generated by a statistical functional

$$(5.10) \qquad T(F) = \int_0^1 h(F^{-1}(x))m(x)dx$$

for any d.f. F .

For a fixed d.f. F , the functional T in equation (5.10) induces a functional τ on $D[0,1]$ defined by

$$\tau(G) = \int_0^1 h(F^{-1}(G^{-1}(x)))m(x)dx$$

for $G \in D[0,1]$. The functional τ is defined for G near the uniform d.f. $U \in D[0,1]$, and we shall show that under appropriate conditions it is Hadamard differentiable at U . Since L-estimators are explicitly defined, no implicit function theorem will be needed, and Hadamard differentiability can be proved directly.

5.3 R-estimators

R-estimators, or rank-estimators, are implicitly defined statistical functionals based on rank statistics. They were introduced by Hodges and Lehmann (1963) and are used to obtain estimates of location in one sample problems and estimates of shift in two sample problems.

To construct an R-estimator of location, we consider a sample $X_1,\ldots,X_n$ with d.f. F and let $R_i^+ = $ rank of $|X_i|$, $i = 1,\ldots,n$ (here we follow Berk (1978)). The rank statistic

$$w_n(X_1,\ldots,X_n) = \sum_{i=1}^{n} J\{(\frac{R_i^+ - \frac{1}{2}}{n})\operatorname{sgn} X_i\}$$

is used to test the null hypothesis that F is symmetric with respect to zero, with the null hypothesis being rejected for large values of w_n. J is a monotone increasing score function defined on $[0,1]$ and is extended to $[-1,1]$ to be odd.

If the population d.f. is symmetric with respect to some parameter θ_o , an estimate θ_n of θ_o is given by a root θ of the equation

$$w_n(X_1-\theta,\ldots,X_n-\theta) = 0 .$$

Since w_n is discontinuous, an exact root might not be achieved, in which case θ_n is the point at which w_n changes sign. In case of multiple roots, these roots will form an interval, and the midpoint of this interval is usually considered to be θ_n .

We can write

$$(\frac{R_i^+ - \frac{1}{2}}{n})\operatorname{sgn} X_i = F_n^*(X_i)-F_n^*(-X_i) , \quad i = 1,\ldots,n ,$$

where

$$F_n^*(x) = \frac{1}{n} \sum_{j=1}^{n} \delta_o^*(x-X_j)$$

with

$$\delta_o^*(x) = \begin{cases} 1 & \text{if } x > 0 \\ \frac{1}{2} & \text{if } x = 0 \\ 0 & \text{if } x < 0 . \end{cases}$$

Let J be extended to $[-1,1]$ to be an odd function, $J(-t) = -J(t)$. Then

$$w_n(X_1,\ldots,X_n) = \int J[F_n^*(x)-F_n^*(-x)]dF_n(x)$$

so

$$(5.11) \qquad w_n(X_1-\theta,\ldots,X_n-\theta) = \int J[F_n^*(x)-F_n^*(2\theta-x)]dF_n(x) .$$

If we use (5.11) as a model, we can define a statistical functional $T(F)$ to be a solution $T(F) = \theta$ of the equation

$$\int J[F(x)-F(2\theta-x)]dF(x) = 0 .$$

For continuous, strictly increasing F , this is equivalent to

$$(5.12) \qquad \int_0^1 J[x-F(2\theta-F^{-1}(x))]dx = 0 .$$

Therefore we define an <u>R-estimator</u> to be a root $T(F_n) = \theta$ of the equation

$$(5.13) \qquad \int_0^1 J[x-F_n(2\theta-F_n^{-1}(x))]dx = 0$$

where F_n is the empirical d.f. corresponding to the sample $X_1,\ldots,X_n$. As before, there may be an interval of roots which satisfy (5.13), in which case usually the midpoint is chosen for $T(F_n)$ (see Hodges and Lehmann (1963)). It should be noted that the roots of (5.11) and (5.13) do not necessarily coincide, but they are asymptotically equal.

When $J(x) = x$ the R-estimator $T(F_n)$ is called the Hodges-Lehmann estimator and $T(F_n) = \frac{1}{2} \underset{i,j}{\mathrm{med}} \{\frac{X_i+X_j}{2}\}$.

Estimates of shift for two independent samples can be obtained in a similar fashion from rank statistics of the same type as above, and the corresponding statistical functional is similar to the estimator defined by equation (5.13). For more details see Huber (1981).

For a fixed d.f. F , the functional T defined by (5.12) will induce a functional τ on $D[0,1]$ defined as a root $\tau(G) = \theta$ of

$$(5.14) \qquad \int_0^1 J[x-G(F(2\theta-F^{-1}(G^{-1}(x))))]dx = 0 ,$$

for $G \in D[0,1]$. The problem of multiple roots remains, and to cope with this we must proceed in a more precise manner.

Let J be continuous, odd, and strictly increasing on $\mathbb{R}$, and suppose that the d.f. F is strictly increasing and absolutely continuous on $\mathbb{R}$. Define $\Phi: D[0,1]\times \mathbb{R} \longrightarrow \mathbb{R}$ by

$$(5.15) \qquad \Phi(G,\theta) = \int_0^1 J[x-G(F(2\theta-F^{-1}(G^{-1}(x))))]dx .$$

We can now define $\tau: D[0,1] \longrightarrow \mathbb{R}$ by

$$\tau(G) = \lambda \sup \{\theta: \Phi(G,\theta) \geq 0\} + (1-\lambda) \inf \{\theta: \Phi(G,\theta) \leq 0\}$$

where λ is fixed in $[0,1]$ and is usually chosen to be $\frac{1}{2}$. Due to the conditions imposed on J , both Φ and τ are defined for all $G \in D[0,1]$.

5.4 Modification of elements of $D[0,1]$

Because of the possibility of multiple roots for equation (5.14), it is inconvenient to deal directly with the function Φ defined in (5.15). To remedy this, we shall introduce in this section a modification procedure for elements of $D[0,1]$ which will transform them into increasing, continuous functions on $[0,1]$. With these modified functions, equations of the form (5.14) will have unique solutions, and will therefore be more tractable mathematically.

__Definition 5.4.1.__ Let $G \in D[0,1]$ and let $0 < \alpha < 1$. Define $\hat{G}$ and $\check{G} \in D[0,1]$ by

$$\hat{G}(x) = \sup_{\substack{0<t\le x \\ x\le s\le 1}} \{G(t) + \alpha(x-t) , G(s) + \alpha^{-1}(x-s)\} ,$$

$$\check{G}(x) = \inf_{\substack{x\le t\le 1 \\ 0\le s\le x}} \{G(t) + \alpha(x-t) , G(s) + \alpha^{-1}(x-s) \} . \qquad \Delta$$

It is easy to see that $\check{G} \le G \le \hat{G}$, that $\hat{G}$ and $\check{G}$ are continuous and strictly increasing, and that $U = \hat{U} = \check{U}$ (recall that $U(x) = x$).

__Lemma 5.4.2.__ If $0 \le a < b \le 1$, then for every $G \in D[0,1]$ we have

$$\alpha(b-a) \le \hat{G}(b)-\hat{G}(a) \le \alpha^{-1}(b-a)$$

$$\alpha(b-a) \le \check{G}(b)-\check{G}(a) \le \alpha^{-1}(b-a) \quad .$$

<u>Proof</u>:

$$\hat{G}(a) = \sup_{\substack{0 \le t \le a \\ a \le s \le 1}} \{G(t) + \alpha(a-t) \, , \, G(s) + \alpha^{-1}(a-s)\}$$

$$(5.16) \qquad \le \sup_{\substack{0 \le t \le b \\ a \le s \le 1}} \{G(t) + \alpha(a-t) \, , \, G(s) + \alpha^{-1}(a-s)\} \, .$$

But $\alpha < \alpha^{-1}$, so for $a \le s \le b$,

$$G(s) + \alpha^{-1}(a-s) \le G(s) + \alpha(a-s)$$

so (5.16) is

$$\le \sup_{\substack{0 \le t \le b \\ b \le s \le 1}} \{G(t) + \alpha(a-t) \, , \, G(s) + \alpha^{-1}(a-s)\}$$

$$= \sup_{\substack{0 \le t \le b \\ b \le s \le 1}} \{G(t) + \alpha(a-b) + \alpha(b-t) \, , \, G(s) + \alpha^{-1}(a-b) + \alpha^{-1}(b-s)\}$$

$$\le \sup_{\substack{0 \le t \le b \\ b \le s \le 1}} \{G(t) + \alpha(b-t) + \alpha(a-b) \, , \, G(s) + \alpha^{-1}(b-s) + \alpha(a-b)\}$$

since $\alpha^{-1}(a-b) \le \alpha(a-b)$. This is

$$= \sup_{\substack{0 \le t \le b \\ b \le s \le 1}} \{G(t) + \alpha(b-t) \, , \, G(s) + \alpha^{-1}(b-s) + \alpha(a-b)\}$$

$$= \hat{G}(b) + \alpha(a-b) \, .$$

Hence $\alpha(b-a) \le \hat{G}(b) - \hat{G}(a)$.

The other inequality for $\hat{G}$ follows in a similar manner, as do the inequalities for $\check{G}$. $\qquad\qquad\qquad \Delta$

The reason for introducing $\hat{G}$ and $\check{G}$ is to smooth the jumps in G, especially when G is an empirical d.f.. The modification in Definition 5.4.1 can be depicted as in Figure 5.1.

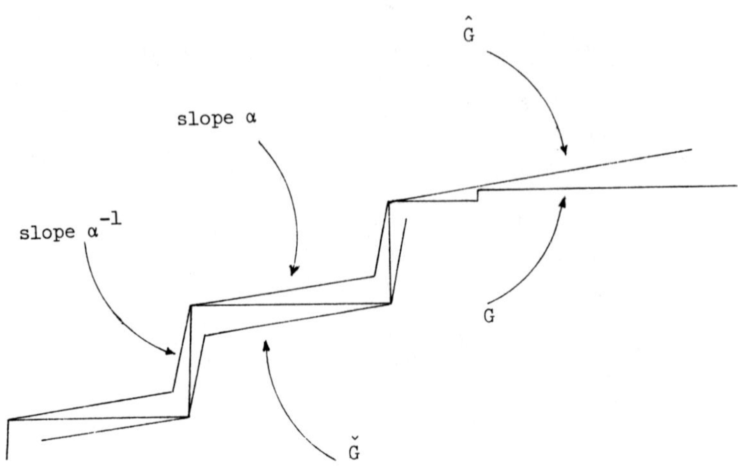

Figure 5.1

Now $\hat{G}$ and $\check{G}$ can be used to define

$$\hat{\Phi}(G,\theta) = \int_0^1 J[x-\hat{G}(F(2\theta-F^{-1}(G^{-1}(x))))]dx \ ,$$

(5.17)

$$\check{\Phi}(G,\theta) = \int_0^1 J[x-\check{G}(F(2\theta-F^{-1}(G^{-1}(x))))]dx \ .$$

Since $\hat{G} \geq G \geq \check{G}$ and J is increasing, it follows immediately that for all θ ,

$$\hat{\Phi}(G,\theta) \leq \Phi(G,\theta) \leq \check{\Phi}(G,\theta) \ .$$

Note that in the particular case that U is the uniform d.f., then $\hat{U} = U = \check{U}$, so

$$\hat{\Phi}(U,\theta) = \Phi(U,\theta) = \check{\Phi}(U,\theta)$$

for all $\theta \in \mathbb{R}$.

<u>Lemma 5.4.3.</u> Let $g: (a,b) \longrightarrow \mathbb{R}$ be continuous and piecewise differen-
tiable. Suppose that

$$- \infty < m = \inf_{t \in (a,b)} g'(t) \ , \qquad \infty > M = \sup_{t \in (a,b)} g'(t) \ .$$

Then for all $x,y \in (a,b)$ with $x \le y$,

$$m(y-x) \le g(y)-g(x) \le M(y-x) \ .$$

<u>Proof</u>: Apply the mean value theorem to each subinterval (x_i, x_{i+1}) and
add the results. Δ

The following theorem will allow us to apply the implicit function
theorem, Theorem 6.2.1, to R-estimators.

<u>Theorem 5.4.4.</u> Let F be a d.f. with continuous, bounded density
$f = F'$ such that $f(\theta_o) > 0$, where θ_o satisfies $\Phi(U,\theta_o) = 0$ for
the uniform d.f. U . Let J be continuous, increasing, odd,
and piecewise differentiable on $\mathbb{R}$. Suppose that there exist numbers
m and M such that $J'(x) \le M < \infty$ for all x for which $J'(x)$
exists and $0 < m \le J'(x)$ for all x in some neighborhood of zero.
Then

    i) for $G,H \in D[0,1]$, $G \ge H$ implies that

$$\hat{\Phi}(G,\theta) \le \hat{\Phi}(H,\theta) \ ,$$

$$\check{\Phi}(G,\theta) \le \check{\Phi}(H,\theta)$$

    for all $\theta \in \mathbb{R}$;

   ii) there exist neighborhoods N_o of θ_o and N_U of U , and
      positive constants A , B such that

58

$$A(\theta-\sigma) \leq \hat{\Phi}(G,\sigma)-\hat{\Phi}(G,\theta) \leq B(\theta-\sigma) \ ,$$

$$A(\theta-\sigma) \leq \check{\Phi}(G,\sigma)-\check{\Phi}(G,\theta) \leq B(\theta-\sigma)$$

for all $\theta,\sigma \in N_o$ with $\theta \geq \sigma$ and $G \in N_U$;

iii) there exists a constant $k > 0$ such that for all G,H in a neighborhood N_U of U ,

$$|\hat{\Phi}(H,\theta)-\hat{\Phi}(G,\theta)| \leq k \ \|G-H\| \ ,$$

$$|\check{\Phi}(H,\theta)-\check{\Phi}(G,\theta)| \leq k \ \|G-H\|$$

for all $\theta \in \mathbb{R}$.

<u>Proof</u>: Choose b such that $f(x) \leq b$ for all $x \in \mathbb{R}$.

(i) If $G \geq H$ then $\hat{G} \geq \hat{H}$ and $G^{-1} \leq H^{-1}$. Since F and F^{-1} are monotone increasing,

$$\hat{G}(F(2\theta-F^{-1}(G^{-1}(x)))) \geq \hat{H}(F(2\theta-F^{-1}(H^{-1}(x))))$$

for all θ . J monotone increasing implies that

$$\hat{\Phi}(G,\theta) \leq \hat{\Phi}(H,\theta) \ .$$

Analogously for $\check{\Phi}$.

(ii) Let $\theta \leq \sigma$ and $G \in D[0,1]$. Then by applying Lemmas 5.4.2 and 5.4.3 we obtain

$$\hat{\Phi}(G,\sigma) - \hat{\Phi}(G,\theta)$$

$$= \int_0^1 \{J[x-\hat{G}(F(2\sigma-F^{-1}(G^{-1}(x))))] - J[x-\hat{G}(F(2\theta-F^{-1}(G^{-1}(x))))]\}dx$$

$$\leq M\int_0^1 [\hat{G}(F(2\theta-F^{-1}(G^{-1}(x)))) - \hat{G}(F(2\sigma-F^{-1}(G^{-1}(x))))]dx$$

$$\le \; M\alpha^{-1} \int_0^1 \{F(2\theta - F^{-1}(G^{-1}(x))) - F(2\sigma - F^{-1}(G^{-1}(x)))\} dx$$

$$\le \; 2bM\alpha^{-1}(\theta - \sigma) \; .$$

If we let $\;B = 2bM\alpha^{-1}\;$, then

$$\hat{\Phi}(G,\sigma) - \hat{\Phi}(G,\theta) \; \le \; B(\theta - \sigma) \; .$$

For the other inequality choose $\;\delta > 0\;$ such that $\;|x| < \delta\;$ implies $J'(x) \ge m > 0\;$. Let $\;x_\theta = \hat{G}(F(2\theta - F^{-1}(G^{-1}(x))))\;$ and let $\;x_o = F(\theta_o)\;$. We shall show that there is an $\;\eta > 0\;$ such that if $\;\theta\;$ is near $\;\theta_o\;$ and $G\;$ is near $\;U\;$, then

$$(5.18) \qquad |x - x_o| < \eta/2 \quad \text{implies} \quad |x - x_\theta| < \delta \; .$$

To prove this, choose $\;\varepsilon > 0\;$ such that $\;\varepsilon < \alpha\delta b^{-1}/2\;$ and $\;f(t) > a > 0\;$ when $\;t \in (\theta_o - \varepsilon, \theta_o + \varepsilon)\;$ (this is possible since $\;f(\theta_o) > 0\;$ and $\;f\;$ is continuous). Then since $\;F^{-1}\;$ is continuous there exists $\;\eta > 0\;$ such that $\;|x - x_o| < \eta\;$ implies

$$|F^{-1}(x) - F^{-1}(x_o)| < \varepsilon/2 \; ,$$

and we can choose $\;\eta < \delta/2\;$. Also if $\;\| G - U \| < \eta/2\;$ then $\| G^{-1} - U \| < \eta/2\;$, so $\;|G^{-1}(x) - x| < \eta/2\;$ and $\;|x - x_o| < \eta/2\;$ implies that $|G^{-1}(x) - x_o| < \eta\;$, and so

$$|F^{-1}(G^{-1}(x)) - F^{-1}(x_o)| < \varepsilon/2 \; .$$

Therefore let $\;\theta \in (\theta_o - \gamma, \theta_o + \gamma)\;$ with $\;\gamma = \varepsilon/4\;$, and let $\;G\;$ be such that $\;\| G - U \| < \eta/2\;$. Then $\;\| G^{-1} - U \| < \eta/2\;$, and if $\;|x - x_o| < \eta/2\;$ we have

$$|x-x_\theta| < \eta/2 + |x_o-x_\theta|$$

$$\leq \eta/2 + |x_o-\hat{G}(x_o)| + |\hat{G}(x_o)-x_\theta|$$

$$\leq \eta/2 + \eta/2 + |\hat{G}(F(2\theta_o-F^{-1}(x_o)))-x_\theta|$$

$$\leq \eta + \alpha^{-1}|F(2\theta_o-F^{-1}(x_o))-F(2\theta-F^{-1}(G^{-1}(x)))|$$

$$\leq \eta + \alpha^{-1} b\{|2\theta_o-2\theta| + |F^{-1}(x_o)-F^{-1}(G^{-1}(x))|\}$$

$$\leq \eta + \alpha^{-1} b\varepsilon/2 + \alpha^{-1} b\varepsilon/2$$

$$\leq \delta$$

so that (5.18) holds.

Now, let $I = \{x: |x-x_o| < \eta\}$ with η small enough such that $I \subset (0,1)$. Then with x_θ as above and x_σ defined in the same manner with σ replacing θ, we have

$$\hat{\Phi}(G,\sigma)-\hat{\Phi}(G,\theta) \geq \int_I [J(x-x_\sigma)-J(x-x_\theta)]dx$$

$$\geq m \int_I (x_\theta-x_\sigma)dx$$

$$\geq m\alpha \int_I \{F(2\theta-F^{-1}(G^{-1}(x)))-F(2\sigma-F^{-1}(G^{-1}(x)))\}dx$$

where θ, σ, η, and G are chosen as above so that $|x-x_\theta| < \delta$ and $|x-x_\sigma| < \delta$. The second inequality follows by the mean value theorem and the fact that $J'(x) \geq m$ for $|x| < \delta$.

By Lemma 5.1.1,

$$F(2\theta - F^{-1}(G^{-1}(x))) - F(2\sigma - F^{-1}(G^{-1}(x)))$$

$$= f[\beta(-F^{-1}(G^{-1}(x))) - F^{-1}(G^{-1}(x))](2\theta - 2\sigma)$$

where $\ 2\sigma \leq \beta(-F^{-1}(G^{-1}(x))) \leq 2\theta$. Therefore

$$\hat{\Phi}(G,\sigma) - \hat{\Phi}(G,\theta)$$

$$\geq m\alpha(2\theta - 2\sigma)\int_I f[\beta(-F^{-1}(G^{-1}(x))) - F^{-1}(G^{-1}(x))]dx \ ,$$

so all that we have to prove is

$$(5.19) \qquad \int_I f[\beta(-F^{-1}(G^{-1}(x))) - F^{-1}(G^{-1}(x))]dx > c > 0$$

for all $\ G \in N_U$, where $\ c\ $ is a constant not depending on $\ G$. This follows if we show that there exists some $\ a > 0\ $ such that

$$(5.20) \qquad f[\beta(-F^{-1}(G^{-1}(x))) - F^{-1}(G^{-1}(x))] \geq a$$

in some subinterval of $\ I\ $ which is independent of $\ G \in N_U$.

Since $\ f(\theta_o) > 0\ $ and $\ f\ $ is continuous, there exists $\ \epsilon > 0\ $ and $\ a > 0\ $ such that $\ \epsilon < \alpha\delta b^{-1}/2\ $ and $\ f(t) \geq a\ $ for all $\ t \in [\theta_o - \epsilon, \theta_o + \epsilon]$. Therefore (5.20) holds when

$$(5.21) \qquad \beta(-F^{-1}(G^{-1}(x))) - F^{-1}(G^{-1}(x)) \in [\theta_o - \epsilon, \theta_o + \epsilon]$$

for $\ \sigma$, $\theta\ $ near $\ \theta_o$, $G\ $ near $\ U$, and $\ x\ $ in some subinterval of $\ I\ $ independent of $\ G$.

Let $\ \gamma = \epsilon/4$. Then since

$$2\sigma \leq \beta(-F^{-1}(G^{-1}(x))) \leq 2\theta \ ,$$

it follows that if $\ \sigma, \theta \in (\theta_o - \gamma, \theta_o + \gamma)\ $ then

$$|\beta(-F^{-1}(G^{-1}(x))) - 2\theta_o| < \epsilon/2 \ .$$

Now F^{-1} is continuous, so there exists $\eta > 0$ with $\eta < \delta/2$ such that $|x-x_o| < \eta$ implies that

$$|F^{-1}(x)-F^{-1}(x_o)| < \varepsilon/2 \ .$$

Let

$$N_U = \{G \in D[0,1]: \|G-U\| < \eta/2\} \ .$$

Simple geometry shows that if $G \in N_U$ then $G^{-1} \in N_U$, so if $|x-x_o| < \eta/2$ and $G \in N_U$, then

$$|G^{-1}(x)-x \ | < \eta/2$$

and

$$|G^{-1}(x)-x_o| < \eta \ .$$

Hence

$$|F^{-1}(G^{-1}(x))-\theta_o| < \varepsilon/2$$

so (5.21) holds, and therefore so do (5.20) and (5.19). Thus

$$\hat{\Phi}(G,\sigma)-\hat{\Phi}(G,\theta) \geq 4m\alpha a\eta(\theta-\sigma)$$

$$= A\ (\theta-\sigma)$$

for $A = 4m\alpha a\eta$.

Analogously for $\check{\Phi}$.

(iii) Let $N_U = \{G: \|G-U\| < \frac{1}{2}\}$, let $k = \sup\limits_{x \in [-1,1]} |J(x)|$, choose $G,H \in N_U$, and let $\varepsilon = \|G-H\|$. Without loss of generality we can assume that $\hat{\Phi}(G,\theta) \geq \hat{\Phi}(H,\theta)$. Since $G + \varepsilon \geq H$,

$$\hat{\Phi}(H,\theta) \geq \hat{\Phi}(G+\varepsilon,\theta)$$

and

$$0 \leq \hat{\Phi}(G,\theta)-\hat{\Phi}(H,\theta) \leq \hat{\Phi}(G,\theta)-\hat{\Phi}(G+\epsilon,\theta) \quad .$$

Now,

$$\hat{\Phi}(G+\epsilon,\theta) = \int_0^1 J[x-(G+\epsilon)^{\hat{}}(F(2\theta-F^{-1}((G+\epsilon)^{-1}(x))))]dx \quad .$$

Since

$$(G+\epsilon)^{-1}(x) = \inf\{1,y: G(y) + \epsilon \geq x\}$$

$$= \inf\{1,y: G(y) \geq x-\epsilon\}$$

$$= G^{-1}(x-\epsilon)$$

and

$$(G+\epsilon)^{\hat{}} = \hat{G} + \epsilon \quad ,$$

we have

$$\hat{\Phi}(G+\epsilon,\theta) = \int_0^1 J[x-\epsilon-\hat{G}(F(2\theta-F^{-1}(G^{-1}(x-\epsilon))))]dx$$

$$= \int_{-\epsilon}^{1-\epsilon} J[z-\hat{G}(F(2\theta-F^{-1}(G^{-1}(z))))]dz$$

$$\geq -2k\epsilon + \hat{\Phi}(G,\theta) \quad .$$

Therefore

$$0 \leq \hat{\Phi}(G,\theta)-\hat{\Phi}(G+\epsilon,\theta) \leq 2k\epsilon$$

so

$$|\hat{\Phi}(G,\theta)-\hat{\Phi}(H,\theta)| < 2k\epsilon \quad .$$

Analogously for $\check{\Phi}$. $\qquad\qquad\qquad\qquad\qquad \Delta$

64

It follows from Theorem 5.4.4, part (ii), that if $\hat{\tau}(G)$ is defined to be a solution of $\hat{\Phi}(G,\theta) = 0$ and $\check{\tau}(G)$ to be a solution of $\check{\Phi}(G,\theta) = 0$, then $\hat{\tau}$ and $\check{\tau}$ are uniquely defined in a neighborhood of U . For G near U we have

$$\hat{\tau}(G) \leq \tau(G) \leq \check{\tau}(G)$$

and

$$\hat{\tau}(U) = \tau(U) = \check{\tau}(U) .$$

These facts will be used in Chapter VII to prove that τ is Hadamard differentiable at U .

CHAPTER VI

CALCULUS ON FUNCTION SPACES

A necessary step in the application of von Mises' method is to
establish that a statistical functional is in some sense differentiable.
To accomplish this, we present here the rudiments of a calculus for
functionals defined on $D[0,1]$, which will enable us to deal with the
statistical functionals that we considered in the last chapter, as well
as others that we shall introduce later.

The functionals that we are interested in can frequently be reduced
to a composition of transformations defined on $C[0,1]$, $D[0,1]$, or
$L^p[0,1]$, $p \geq 1$. If a functional can be expressed as a composition of
transformations all of which are Hadamard (or Fréchet) differentiable,
then the chain rule implies that the functional itself will be Hadamard
differentiable. Our strategy, then, is to establish the differentiability of certain basic types of transformations, and then to decompose more
complex transformations into a composition of these basic ones, just as
is done in common differential calculus. Although we cannot provide here
an exhaustive collection of differentiable transformations, the ones
that we do present are representative of the types of transformations
which commonly occur in nonlinear statistical functionals.

Throughout this chapter, for a composition $H \circ G$ of real valued
functions of a real variable, if $G(t)$ lies outside the domain of H ,
then $H(G(t))$ is defined to be $H(x)$ where x is the point in the

domain of H nearest $G(t)$. U will represent the function $U(x) = x$, $x \in [0,1]$. The spaces $C[0,1]$ and $D[0,1]$ will have the uniform topology and $L^p[0,1]$, $p \geq 1$, will have the usual L^p topology.

6.1 Differentiability theorems

The relation $G \longrightarrow G^{-1}$ which associates to each element $G \in D[0,1]$ the inverse function defined by $G^{-1}(y) = \inf \{1,x: G(x) \geq y\}$ defines a transformation $\xi: D[0,1] \longrightarrow D[0,1]$. Example 2.3.2 precludes ξ from being Fréchet differentiable at U and it is not hard to show that ξ is not even Gateaux differentiable at U . Consider for example, $H(x) = 1$ for all $x \in [0,1]$. Then for

$$\mathrm{Rem}(tH) = \xi(U+tH) - \xi(U) - \xi'_U(tH) .$$

where

$$\xi'_U(tH) = -tH$$

(this derivative appears below in Proposition 6.1.1), we have

$$\mathrm{Rem}(tH)/t = \begin{cases} 1-y/t & \text{for } 0 \leq y \leq t \\ 0 & \text{otherwise} \end{cases}$$

Therefore $\|\mathrm{Rem}(tH)/t\| = 1$ so $\mathrm{Rem}(tH)/t$ does not tend to zero in $D[0,1]$.

However if $L^p[0,1]$, $p \geq 1$, is the codomain of ξ , then ξ will be Hadamard differentiable as is shown in the next proposition due to Reeds (1976).

<u>Proposition 6.1.1.</u> Let $\xi: D[0,1] \longrightarrow L^p[0,1]$, $p \geq 1$, be the mapping $\xi(G) = G^{-1}$. Then ξ is Hadamard differentiable at U with derivative

$$\xi'_U(H) = -H .$$

Proof: We shall first prove that ξ is continuous at U. Let $\epsilon > 0$. If $\|G-U\| < \epsilon$, then $\|G^{-1}-U\| < \epsilon$, so $G \longrightarrow G^{-1}$ is continuous as a transformation from $D[0,1]$ to $D[0,1]$. Since the inclusion $D[0,1] \longrightarrow L^p[0,1]$, $p \geq 1$, is continuous, ξ is continuous at U.

To prove the differentiability of ξ at U, let $K \subset D[0,1]$ be compact. Let $H \in K$ and choose k such that $\|H\| < k$ for all $H \in K$. We must prove that

$$\frac{\mathrm{Rem}(tH)}{t} = \frac{(U+tH)^{-1}-U+tH}{t} \xrightarrow{L^p} 0$$

uniformly for $H \in K$ as $t \longrightarrow 0$. For fixed $t \neq 0$, let $G = U + tH$ and for each $y \in [0,1]$ define $x^+ = G^{-1}(y)$ and $x^- = G^{-1}(y)-t^2$. Then

$$y \leq G(x^+)$$

by definition of G^{-1} and right continuity of G. Therefore

$$y \leq U(x^+) + tH(x^+)$$

$$= x^+ + tH(x^+)$$

$$= G^{-1}(y) + tH(x^+)$$

$$< G^{-1}(y) + tH(x^+) + t^2$$

so

(6.1) $$y-G^{-1}(y) < tH(x^+) + t^2 .$$

Also

$$y > G(x^-)$$

by definition of G^{-1} , so

$$y > x^- + tH(x^-)$$

$$= G^{-1}(y) + tH(x^-) - t^2$$

and

(6.2) $$y - G^{-1}(y) > tH(x^-) - t^2 \ .$$

From (6.1) and (6.2) it follows that

$$-t^2 + t[H(x^-)-H(y)] \le y - G^{-1}(y) - tH(y) \le t^2 + t[H(x^+)-H(y)] \ .$$

Let $\alpha(t,H,y) = \dfrac{[G^{-1}(y) - y + tH(y)]}{t}$, then

$$|\alpha(t,H,y)| \le |t| + |H(y)-H(x^-)| + |H(y)-H(x^+)|$$

so

$$[\int_0^1 |\alpha(t,H,y)|^p dy]^{1/p} \le |t| + [\int_0^1 |H(y)-H(x^-)|^p dy]^{1/p}$$

$$+ [\int_0^1 |H(y)-H(x^+)|^p dy]^{1/p} \ .$$

Therefore, since $\alpha(t,H,\cdot) = \mathrm{Rem}(tH)/t$, it suffices to prove that

(6.3) $$\int_0^1 |H(y)-H(x^+)|^p dy \longrightarrow 0$$

and

$$\int_0^1 |H(y)-H(x^-)|^p dy \longrightarrow 0$$

uniformly for $H \in K$ as $t \longrightarrow 0$.

Now (6.1) and (6.2) imply that

$$tH(x^-) - t^2 < y-x^+ < tH(x^+) + t^2$$

and

$$tH(x^-) < y - x^- < tH(x^+) + 2t^2 .$$

Hence

$$|x^+-y| \leq |t|k + t^2$$

and

$$|x^--y| \leq |t|k + 2t^2 .$$

Therefore if (a,b) is an interval in $[0,1]$, we can choose t small enough such that if $y \in (a+2|t|k,b-2|t|k)$, then $x^-,x^+ \in (a,b)$.

Let $\varepsilon > 0$. By Proposition 4.1.3 there exists a partition $0 = a_0 < a_1 < \ldots < a_{n(\varepsilon)} = 1$ of $[0,1]$ and points $p_j \in [a_{j-1},a_j)$ such that for all $H \in K$,

$$|H(x)-H(p_j)| < \varepsilon \quad \text{for all } x \in [a_{j-1},a_j) .$$

Let

$$I_j = [a_{j-1},a_j)$$

and let

$$\tilde{I}_j = (a_{j-1}+2|t|k,a_j-2|t|k)$$

for all $j = 1,\ldots,n(\varepsilon)$. Then for t small enough,

$$\int_{I_j} |H(y)-H(x^+)|^p dy = \int_{\tilde{I}_j} |H(y)-H(x^+)|^p + \int_{I_j-\tilde{I}_j} |H(y)-H(x^+)|^p dy$$

$$\leq (2\varepsilon)^p(a_j-a_{j-1}) + 8k^{p+1}|t|$$

so

$$\sup_{H \in K} \int_0^1 |H(y) - H(x^+)|^p dy \leq (2\epsilon)^p + n(\epsilon) \; 8k^{p+1}|t| \; .$$

Since $\epsilon > 0$ is arbitrary, it follows that

$$\int_0^1 |H(y) - H(x^+)|^p dy \longrightarrow 0$$

uniformly for $H \in K$ as $t \longrightarrow 0$. An analogous argument shows that

$$\int_0^1 |H(y) - H(x^-)|^p dy \longrightarrow 0$$

uniformly for $H \in K$ as $t \longrightarrow 0$. Δ

We shall need several results dealing with the differentiability of the composition of functions. The first such result is

<u>Proposition 6.1.2.</u> Let $L: \mathbb{R} \longrightarrow \mathbb{R}$ be continuous and piecewise differentiable with bounded derivative. Let $\gamma: L^p[0,1] \longrightarrow L^p[0,1]$, $p \geq 1$, be defined by $\gamma(S) = L \circ S$. Let A be the set of points in $\mathbb{R}$ where L is not differentiable. If L is defined in a neighborhood of $S \in L^p[0,1]$ and if $\mu\{x: S(x) \in A\} = 0$, where μ is Lebesgue measure, then γ is Hadamard differentiable at S with derivative

$$\gamma_S'(H) = (L' \circ S)H \; .$$

<u>Proof:</u> For a compact set $K \subset L^p[0,1]$, we must show that

$$\left\| \frac{Rem(th)}{t} \right\|_p \longrightarrow 0$$

uniformly for $H \in K$, as $t \longrightarrow 0$, where

$$Rem(tH) = L \circ (S+tH) - L \circ S - (L' \circ S)tH \; .$$

Since K is compact, for any $\varepsilon > 0$ we can choose $H_1,\ldots,H_n \in K$ such that

$$\inf_{1 \le i \le n} \| H - H_i \|_p < \varepsilon$$

for any $H \in K$. Since for a given H_i

$$\frac{\mathrm{Rem}(tH_i)}{t} = \frac{L(S(x)+tH_i(x)) - L(S(x))}{t} - L'(S(x))H_i(x) \; ,$$

by Lemma 5.4.3 we have

$$\left| \frac{L(S(x)+tH_i(x)) - L(S(x))}{t} \right| \le M|H_i(x)| \; ,$$

where M is a bound for $|L'|$. Therefore for each $i = 1,\ldots,n$,

$$\left| \frac{\mathrm{Rem}(tH_i)(x)}{t} \right| \le 2M|H_i(x)| .$$

Moreover, for x such that $S(x) \notin A$,

$$\frac{\mathrm{Rem}(tH_i)(x)}{t} \longrightarrow 0 \; ,$$

as $t \longrightarrow 0$, so by dominated convergence,

$$(6.4) \qquad \left\| \frac{\mathrm{Rem}(tH_i)}{t} \right\|_p \longrightarrow 0 \; .$$

For any $H \in K$,

$$(6.5) \qquad \left\| \frac{\mathrm{Rem}(tH)}{t} \right\|_p \le \left\| \frac{\mathrm{Rem}(tH_i)}{t} \right\|_p + \left\| \frac{\mathrm{Rem}(tH)}{t} - \frac{\mathrm{Rem}(tH_i)}{t} \right\|_p \; .$$

Also,

$$\frac{\text{Rem}(tH)}{t} - \frac{\text{Rem}(tH_i)}{t}$$

$$= \frac{L(S(x)+tH(x)) - L(S(x)+tH_i(x))}{t} - L'(S(x))[H(x)-H_i(x)] \ .$$

By Lemma 5.4.3 we have

$$\frac{L(S(x)+tH(x)) - L(S(x)+tH_i(x))}{t} \leq M|H(x)-H_i(x)|$$

so

$$\inf_{1 \leq i \leq n} \left\|\frac{\text{Rem}(tH)}{t} - \frac{\text{Rem}(tH_i)}{t}\right\|_p \leq 2M\epsilon \ .$$

Since $\epsilon > 0$ is arbitrary and M is fixed, this last inequality together with (6.4) and (6.5) implies that

$$\left\|\frac{\text{Rem}(tH)}{t}\right\|_p \longrightarrow 0$$

uniformly for $H \in K$, as $t \longrightarrow 0$. $\qquad\qquad\qquad \Delta$

The following proposition is similar to the one we just proved, but includes an independent parameter in the composition of functions.

<u>Proposition 6.1.3.</u> Let $\Gamma : \mathbb{R} \times \mathbb{R} \longrightarrow \mathbb{R}$ be a continuously differentiable function such that the partial derivatives $\Gamma_1(x,\theta)$ and $\Gamma_2(x,\theta)$ are bounded and uniformly continuous in x for $\theta = \theta_o$, and suppose that $\Gamma_1(x,\theta) \longrightarrow \Gamma_1(x,\theta_o)$ uniformly in x as $\theta \longrightarrow \theta_o$. Let $\gamma : L^{2p}[0,1] \times \mathbb{R} \longrightarrow L^p[0,1]$, $p \geq 1$, be defined by $\gamma(S,\theta)(x) = \Gamma(S(x),\theta)$. Then γ is Fréchet differentiable at (S,θ_o) , $S \in L^{2p}[0,1]$, with Taylor expansion

$$\gamma(S+tH,\theta_o+th) = \gamma(S,\theta_o) + t\Gamma_1(S,\theta_o)H + t\Gamma_2(S,\theta_o)h + \text{Rem}(tH,th) \ .$$

<u>Proof</u>: Let $B \subset L^2[0,1]$ be bounded and let $k > 0$. For $H \in B$, $|h| < k$,

$$\frac{\text{Rem}(tH,th)(x)}{t} = \frac{\gamma(S+tH,\theta_o+th)(x) - \gamma(S,\theta_o)(x)}{t}$$

$$- \Gamma_1(S(x),\theta_o)H(x) - \Gamma_2(S(x),\theta_o)h$$

$$= \frac{\Gamma(S(x)+tH(x),\theta_o+th) - \Gamma(S(x)+tH(x),\theta_o)}{t} - \Gamma_2(S(x),\theta_o)h$$

$$+ \frac{\Gamma(S(x)+tH(x),\theta_o) - \Gamma(S(x),\theta_o)}{t} - \Gamma_1(S(x),\theta_o)H(x)$$

$$= \Gamma_2(S(x)+tH(x),\theta_o+\beta(x,t))h - \Gamma_2(S(x),\theta_o)h$$

$$+ \Gamma_1(S(x)+\delta(x,t),\theta_o)H(x) - \Gamma_1(S(x),\theta_o)H(x)$$

where the existence of functions β and δ satisfying $|\beta(x,t)| < |th|$ and $|\delta(x,t)| < |tH(x)|$ is implied by the mean value theorem.

Therefore,

$$(6.6) \qquad \left\| \frac{\text{Rem}(tH,th)}{t} \right\|_p$$

$$\leq |h| \left[\int_0^1 |\Gamma_2(S(x)+tH(x),\theta_o+\beta(x,t)) - \Gamma_2(S(x)+tH(x),\theta_o)|^p dx \right]^{1/p}$$

$$+ |h| \left[\int_0^1 |\Gamma_2(S(x)+tH(x),\theta_o) - \Gamma_2(S(x),\theta_o)|^p dx \right]^{1/p}$$

$$+ \left[\int_0^1 |H(x)|^p |\Gamma_1(S(x)+\delta(x,t),\theta_o) - \Gamma_1(S(x),\theta_o)|^p dx \right]^{1/p} \ .$$

Since $\Gamma_2(x,\theta) \longrightarrow \Gamma_2(x,\theta_o)$ uniformly in x as $\theta \longrightarrow \theta_o$, it follows that the first term on the right hand side of (6.6) tends to zero uniformly for $H \in B$ as $t \longrightarrow 0$.

Define

$$A_t(H) = \{x \in [0,1]: |H(x)| > t^{-\frac{1}{2}}\} .$$

Then

$$\sup_{H \in B} \mu(A_t(H)) \longrightarrow 0 \quad \text{as} \quad t \longrightarrow 0$$

where μ is Lebesgue measure on $[0,1]$. For the second term on the right hand side of (6.6) we have

$$(6.7) \qquad \int_0^1 |\ldots|^p dx = \int_{A_t(H)} |\ldots|^p dx + \int_{A_t^c} |\ldots|^p dx$$

where $A_t^c = [0,1] - A_t(H)$ and

$$|\ldots| = |\Gamma_2(S(x) + tH(x), \theta_o) - \Gamma_2(S(x), \theta_o)| .$$

If $|\Gamma_2| \le M$, then

$$\int_{A_t(H)} |\ldots|^p dx \le 2M^p \mu(A_t(H))$$

which tends to zero uniformly for $H \in B$ as $t \longrightarrow 0$. Since Γ_2 is uniformly continuous in x for $\theta = \theta_o$,

$$\int_{A_t^c} |\ldots|^p dx \longrightarrow 0$$

uniformly for $H \in B$ as $t \longrightarrow 0$.

The third term on the right hand side of (6.6) is

$$(6.8) \qquad \le \|H\|_{2p} \left[\int_0^1 |\Gamma_1(S(x) + \delta(x,t), \theta_o) - \Gamma_1(S(x), \theta_o)|^{2p} dx\right]^{1/2p}$$

by the Cauchy-Schwarz inequality. The integral in (6.8) tends to zero
uniformly for $H \in B$ as $t \longrightarrow 0$ using the same reasoning as was used
for (6.7).

$$\Delta$$

The principal application of this proposition will be in situations
where Γ is derived from a d.f. on $\mathbb{R}$. Let F be a d.f. and let

$$\Gamma(x,\theta) = F(2\theta - F^{-1}(x)) \ .$$

In this case Γ is defined on $[0,1] \times \mathbb{R}$, but can be extended to
$\mathbb{R} \times \mathbb{R}$ by

$$\Gamma(x,\theta) = 1-x \quad \text{for} \quad x \notin [0,1] \ .$$

To apply Proposition 6.1.3 we must show that Γ_1 and Γ_2 satisfy the
hypotheses with

$$\Gamma_1(x,\theta) = - \frac{F'(2\theta - F^{-1}(x))}{F'(F^{-1}(x))}$$

and

$$\Gamma_2(x,\theta) = 2F'(2\theta - F^{-1}(x)) \ .$$

We are interested, in particular, in the case where F is symmetric
about θ_0 . We shall say that the density F' is <u>regular</u> if it has
limits at $+\infty$ and $-\infty$.

<u>Corollary 6.1.4.</u> Let F be a d.f. on $\mathbb{R}$ which is symmetric about θ_0
and assume that F has continuous, regular density F' satisfying
$0 < F' \leq b < \infty$. Then the transformation $\gamma: L^{2p}[0,1] \times \mathbb{R} \longrightarrow L^{p}[0,1]$
defined by

$$\gamma(S,\theta) = F(2\theta - F^{-1}(S))$$

is Fréchet differentiable at (S,θ_0) , $S \in L^{2p}[0,1]$.

76

<u>Proof</u>: Since $F'(2\theta_o - x) = F'(x)$, it follows that $\Gamma_1(x, \theta_o) = -1$ and is therefore uniformly continuous in x . Likewise

$$\Gamma_2(x, \theta_o) = 2F'(2\theta_o - F^{-1}(x))$$

$$= 2F'(F^{-1}(x))$$

is continuous in x and equal to zero outside $[0,1]$. Therefore it is uniformly continuous in x for $\theta = \theta_o$.

The fact that F' is regular and continuous implies that it is uniformly continuous, so $\Gamma_2(x, \theta_o)$ is uniformly continuous in x . Therefore Proposition 6.1.3 can be applied. $\qquad\qquad \Delta$

For the next theorem on the differentiability of the composition of functions, we shall need

<u>Lemma 6.1.5</u>. Let $K \subset D[0,1]$ be compact and $B \subset L^p[0,1]$ be bounded. Then

$$\sup_{\substack{H \in K \\ \alpha, \beta \in B}} \left[\int_0^1 |H(x + t\alpha(x)) - H(x + t\beta(x))|^p dx \right]^{1/p} \longrightarrow 0$$

as $t \longrightarrow 0$. (Here we have extended H to be constant on $(-\infty, 0]$ and $[1, \infty)$, as usual.)

<u>Proof</u>: Let $\varepsilon > 0$ and choose a partition $0 = x_o < x_1 < \ldots < x_n = 1$ of $[0,1]$ with the properties given in Proposition 4.1.3. Let $\delta > 0$ be small enough such that $2n\delta < \varepsilon$. Let $I_j = [x_{j-1}, x_j)$, $j = 1, \ldots, n$ and define $\tilde{I}_j = [x_{j-1} + \delta, x_j - \delta)$, $j = 1, \ldots, n$. Make δ smaller, if necessary, to ensure that $\tilde{I}_j \neq \emptyset$, $j = 1, \ldots, n$.

We have

$$\left[\int_0^1 |H(x+t\alpha(x))-H(x+t\beta(x))|^p dx\right]^{1/p}$$

$$\leq \left[\int_0^1 |H(x+t\alpha(x))-H(x)|^p dx\right]^{1/p} + \left[\int_0^1 |H(x+t\beta(x))-H(x)|^p dx\right]^{1/p} .$$

For $\alpha,\beta \in B$, let

$$A_t = \{x \in [0,1]: |t\alpha(x)| \leq \delta , |t\beta(x)| \leq \delta\} .$$

For $A_t^c = [0,1]-A_t$, $\mu(A_t^c) \longrightarrow 0$ uniformly for $\alpha,\beta \in B$ as $t \longrightarrow 0$, where μ is Lebesgue measure. Now,

$$\int_0^1 |H(x+t\alpha(x)) - H(x)|^p dx$$

$$\leq \sum_{j=1}^n \int_{\tilde{I}_j \cap A_t} |\ldots|^p + \sum_{j=1}^n \int_{I_j-\tilde{I}_j} |\ldots|^p dx + \int_{A_t^c} |\ldots|^p dx$$

(where $|\ldots| = |H(x+t\alpha(x)) - H(x)|$)

$$\leq (2\epsilon)^p + (2\|H\|)^p 2n\delta + (2\|H\|)^p \mu(A_t^c)$$

$$\leq (2\epsilon)^p + 2^p k\epsilon + 2^p k\mu(A_t^c)$$

where $k = \sup \{\|H\|^p: H \in K\}$. Since $\epsilon > 0$ was arbitrary and $\mu(A_t^c) \longrightarrow 0$ as $t \longrightarrow 0$, and since we can apply the same reasoning with β replacing α , the lemma is proved. $\triangle$

In the following proposition we consider the composition of functions where both functions are variable.

<u>Proposition 6.1.6.</u> Let $\Psi: D[0,1] \times L^p[0,1] \longrightarrow L^p[0,1]$ be defined by $\Psi(G,Q) = G\circ Q$. Assume that Q is representable by a differentiable function with range $[0,1]$ and derivative bounded away from zero and infinity. Then Ψ is Hadamard differentiable at (U,Q) , where

$U(x) = x$, with Taylor expansion

$$\Psi(U+tH_1, Q+tH_2) = Q + tH_1 \circ Q + tH_2 + \mathrm{Rem}(tH_1, tH_2) \ .$$

<u>Proof</u>: Let $K_1 \subset D[0,1]$ and $K_2 \subset L^p[0,1]$ be compact subsets. For $H_1 \in K_1$, $H_2 \in K_2$ we have

$$\Psi(U+tH_1, Q+tH_2)(x)$$

$$= (U+tH_1) \circ (Q+tH_2)(x)$$

$$= Q(x) + t[H_1 \circ Q(x) + H_2(x)]$$

$$+ t[H_1 \circ (Q+tH_2)(x) - H_1 \circ Q(x)] \ .$$

Then $\mathrm{Rem}(tH_1, tH_2)(x) = t[H_1 \circ (Q+tH_2)(x) - H_1 \circ Q(x)]$ and

$$\left\| \frac{\mathrm{Rem}(tH_1, tH_2)}{t} \right\|_p = [\int_0^1 |H_1(Q(x)+tH_2(x)) - H_1(Q(x))|^p dx]^{1/p}$$

$$(6.9) \qquad = [\int_{y_0}^{y_1} |H_1(y+tH_2(Q^{-1}(y))) - H_1(y)|^p \frac{dQ^{-1}(y)}{dy} \, dy]^{1/p} \ ,$$

where $y_0 = Q^{-1}(0)$ and $y_1 = Q^{-1}(1)$.

By hypothesis

$$\left| \frac{dQ^{-1}(y)}{dy} \right| \leq M^p$$

for some finite M , so (6.9) is

$$(6.10) \qquad \leq M \, [\int_{y_0}^{y_1} |H_1(y+tH_2(Q^{-1}(y))) - H_1(y)|^p dy]^{1/p} \ .$$

But

$$[\int_{y_0}^{y_1} |H_2(Q^{-1}(y))|^p dy]^{1/p} \leq [\int_0^1 |H_2(x)|^p |Q'(x)| dx]^{1/p}$$

$$\leq \sup_{x \in [0,1]} |Q'(x)|^{1/p} \|H_2\|_p$$

$$< \infty \quad .$$

Hence we can apply Lemma 6.1.5 and the integral in (6.10) tends to zero

uniformly for $H_1 \in K_1$. $\triangle$

The next proposition treats the case where the functions in

Proposition 6.1.6 have been modified according to Definition 5.4.1.

This version will be used to prove differentiability of the functions

$\hat{\Phi}$ and $\check{\Phi}$ defined in Chapter V. Recall that these functions were con-

structed in such a way that an implicit function theorem could be

applied.

<u>Proposition 6.1.7</u>. Under the hypothesis of Proposition 6.1.6, the

transformations

$$\Psi_i : D[0,1] \times L^p[0,1] \longrightarrow L^p[0,1] , \quad i = 1,2 ,$$

defined by

$$\Psi_1(G,Q) = \hat{G} \circ Q \quad \text{and} \quad \Psi_2(G,Q) = \check{G} \circ Q ,$$

are Hadamard differentiable at (U,Q) with the same Taylor expansion as

Ψ .

<u>Proof</u>: Because of Proposition 6.1.6, it suffices to prove that for com-

pact subsets $K_1 \subset D[0,1]$ and $K_2 \subset L^p[0,1]$,

$$(6.11) \qquad \int_0^1 \left| \frac{(U+tH_1)\hat{\ }\circ(Q+tH_2)(x) - (U+tH_1)\circ(Q+tH_2)(x)}{t} \right|^p dx \longrightarrow 0$$

uniformly for $H_1 \in K_1$ and $H_2 \in K_2$ as $t \longrightarrow 0$. Now,

$$0 \leq (U+tH_1)\hat{\ }\circ(Q+tH_2)(x) - (U+tH_1)\circ(Q+tH_2)(x) \qquad \text{a.e.}$$

$$= (U+tH_1)\hat{\ }(z) - (U+tH_1)(z)$$

where $z = (Q+tH_1)(z)$ a.e.. By the definition of $\hat{G}$ (see Definition 5.4.1) there are two possibilities:

(i) There exists $z' \leq z$ such that

$$(U+tH_1)\hat{\ }(z) \leq (U+tH_1)(z') + \alpha(z-z') + t^2 ,$$

so

$$0 \leq (U+tH_1)\hat{\ }(z) - (U+tH_1)(z)$$

$$\leq (U+tH_1)(z') + \alpha(z-z') + t^2 - (U+tH_1)(z)$$

$$(6.12)$$

$$= (1-\alpha)(z-z') + t[H_1(z')-H_1(z)] + t^2$$

$$\leq t[H_1(z')-H_1(z)] + t^2$$

since $0 < \alpha < 1$.

ii) There exists $z' \geq z$ such that

$$(U+tH_1)\hat{\ }(z) \leq (U+tH_1)(z') + \alpha^{-1}(z-z') + t^2 ,$$

so

$$0 \leq (U+tH_1)\hat{\ }(z) - (U+tH_1)(z)$$

$$\leq (U+tH_1)(z') + \alpha^{-1}(z-z') + t^2 - (U+tH_1)(z)$$

$$(6.13) \qquad = (\alpha^{-1}-1)(z-z') + t[H_1(z')-H_1(z)] + t^2$$

$$\leq t[H_1(z')-H_1(z)] + t^2 \ .$$

In both cases, $|z'-z| \leq 2tk/(1-\alpha)$, where $k = \sup \{\|H_1\|: H_1 \in K_1\}$.

Also, in both cases z' depends (Borel) measurably on z since we can

choose z' to be the nearest point above or below z such that

$(U+tH_1)^{\wedge}(z') = (U+tH_1)(z)$.

Since $z = (Q+tH_2)(x)$ a.e., we can write $z' = (Q+tH_3)(x)$ for

some $H_3 \in L^p[0,1]$. Then

$$|H_2(x)-H_3(x)| \leq 2k/(1-\alpha) \qquad \text{a.e.}$$

so H_3 is in a bounded set of $L^p[0,1]$. In view of (6.12) and (6.13)

it suffices to prove that

$$(6.14) \qquad \int_0^1 |H_1(Q(x)+tH_3(x)) - H_1(Q(x)+tH_2(x))|^p dx \longrightarrow 0$$

uniformly in H_1 , H_2 , and H_3 as $t \longrightarrow 0$. To prove this,

Lemma 6.1.5 can be applied as it was for the similar integral in (6.10)

of Proposition 6.1.6. Therefore (6.14) holds.

The proof of $\check{G}$ is analogous. $\qquad\qquad\qquad\qquad\qquad\qquad \Delta$

The next proposition shows that integration with a variable limit

is Hadamard differentiable.

<u>Proposition 6.1.8.</u> Let $\Phi : \mathbb{R} \times L^1(\mathbb{R}) \longrightarrow \mathbb{R}$ be defined by

$$\Phi(m,G) = \int_{-\infty}^m G(x)dx \ .$$

Suppose that G can be represented by a function which is continuous at m. Then Φ is Hadamard differentiable at (m,G) with derivative

$$\Phi'_{(m,G)}(h,H) = hG(m) + \int_{-\infty}^{m} H(x)dx .$$

<u>Proof</u>: Let $K \subset L^1(\mathbb{R})$ be compact and let $k > 0$. Then for $H \in K$ and $|h| < k$,

$$\mathrm{Rem}(t,h,H) = \Phi(m+th,G+tH) - \Phi(m,G) - thG(m) - t\int_0^m H(x)dx$$

$$= \int_{-\infty}^{m+th} (G+tH)(x)dx - \int_{-\infty}^{m} G(x)dx - thG(m) - t\int_0^m H(x)dx$$

$$= \int_m^{m+th} G(x)dx - thG(m) + t\int_m^{m+th} H(x)dx .$$

For $\varepsilon > 0$, if t is small enough, then for $|x-m| < t|h|$ we have $|G(m)-G(x)| < \varepsilon$ by the continuity of G at m, so

$$\left| \int_m^{m+th} G(x)dx - thG(m) \right| < \varepsilon t|h| .$$

Since K is compact, there are $H_1,\ldots,H_n \in K$ such that

$$\inf_i \|H_i - H\| < \varepsilon \text{ for all } H \in K .$$

By bounded convergence,

$$\int_m^{m+th} |H_i(x)|dx \longrightarrow 0 \quad \text{for} \quad i = 1,\ldots,n$$

uniformly for $|h| < k$ as $t \longrightarrow 0$. Therefore, for t small enough,

$$\sup_{i} \int_{m}^{m+th} |H_i(x)| \, dx < \varepsilon \, .$$

Hence

$$\left| \int_{m}^{m+th} H(x) \, dx \right| \leq \| H - H_i \| + \int_{m}^{m+th} |H_i(x)| \, dx$$

$$< 2\varepsilon$$

so

$$\mathrm{Rem}(t,h,H) = o(t)$$

uniformly for $|h| < k$ and $H \in K$. Δ

6.2 An implicit function theorem for statistical functionals

In this section we prove an implicit function theorem which is directly applicable to implicitly defined statistical functionals. Let V represent either $D[0,1]$ or $C[0,1]$ with the uniform topology.

<u>Theorem 6.2.1</u> Let $\Psi: V \times \mathbb{R} \longrightarrow \mathbb{R}$ be Hadamard differentiable at (G_o, θ_o) , suppose that $\Psi(G_o, \theta_o) = 0$, and suppose that there exist neighborhoods N_o of θ_o and M_o of G_o such that:

(i) there exist positive constants A , B   such that if

$\theta, \sigma \in N_o$, $\sigma \leq \theta$ and $G \in M_o$ then

$$A(\theta - \sigma) \leq \Psi(G,\sigma) - \Psi(G,\theta) \leq B(\theta - \sigma) \, ;$$

(ii) there is a positive constant k   such that for all $H, G \in M_o$ and for all $\theta \in N_o$

$$|\Psi(H,\theta) - \Psi(G,\theta)| \leq k \|G - H\| \, .$$

Then there exist a neighborhood M of G_o and a continuous functional $\tau: M \longrightarrow \mathbb{R}$ such that for all $G \in M$,

$$\Psi(G, \tau(G)) = 0 ,$$

and τ is Hadamard differentiable at G_o with derivative

$$\tau'_{G_o} = -(D_2\Psi(G_o, \theta_o))^{-1} \circ D_1\Psi(G_o, \theta_o) .$$

<u>Proof</u>: Choose $\varepsilon > 0$ such that $[\theta_o - \varepsilon, \theta_o + \varepsilon] \subset N_o$. Define

$$M = \{G \in M_o: \|G - G_o\| < A\varepsilon/2k\}$$

and

$$N = \{t \in \mathbb{R}: |t| < A\varepsilon/2\} .$$

Then for $G \in M$ and $t \in N$ we have

(6.15) $$|\Psi(G, \theta_o) - \Psi(G_o, \theta_o)| + |t| \leq A\varepsilon$$

by (ii). Consider

(6.16) $$\Psi(G, \theta) - t = [\Psi(G, \theta) - \Psi(G, \theta_o)] + [(\Psi(G, \theta_o) - \Psi(G_o, \theta_o)) - t]$$

and note that for $\theta = \theta_o + \varepsilon$, the first bracket in (6.16) is equal to

(6.17) $$\Psi(G, \theta_o + \varepsilon) - \Psi(G, \theta_o) \leq -A\varepsilon$$

by (i), whereas for $\theta = \theta_o - \varepsilon$, it is equal to

(6.18) $$\Psi(G, \theta_o - \varepsilon) - \Psi(G, \theta_o) \geq A\varepsilon .$$

Hence equations (6.15), (6.16), (6.17), and (6.18) imply that $\Psi(G, \theta) - t$ takes on both positive and negative values for $\theta \in [\theta_o - \varepsilon, \theta_o + \varepsilon] \subset N_o$.

Now, $\Psi(G,\theta)-t$ is a continuous function of θ , by (i), so there exists some $\theta \in [\theta_o-\varepsilon,\theta_o+\varepsilon]$ such that

$$(6.19) \qquad\qquad \Psi(G,\theta)-t = 0 \ .$$

This root θ is unique since $\Psi(G,\theta)$ is strictly decreasing in θ .

Let $T(G,t)$ be the unique solution of (6.19). Then $T: M \times N \longrightarrow \mathbb{R}$ and satisfies

$$\Psi(G,T(G,t)) = t$$
$$(6.20)$$
$$T(G,\Psi(G,\theta)) = \theta$$

for $\theta \in [\theta_o-\varepsilon,\theta_o+\varepsilon]$. This function T has the following properties:

(i*) for $t,s \in N$, $s \le t$, and for $G \in M$

$$B^{-1}(t-s) \le T(G,s)-T(G,t) \le A^{-1}(t-s) \ ;$$

(ii*) for $G,H \in M$, $t \in N$,

$$|T(G,t)-T(H,t)| \le kA^{-1}\|G-H\| \ .$$

To prove (i*), note that by (i) we have

$$A(T(G,s)-T(G,t)) \le \Psi(G,T(G,t)) - \Psi(G,T(G,s)) = (t-s)$$

and

$$B(T(G,s)-T(G,t)) \ge \Psi(G,T(G,t)) - \Psi(G,T(G,s)) = (t-s) \ .$$

To prove (ii*) we can assume without loss of generality that $T(G,t) \ge T(H,t)$. Then

$$0 \le T(G,t)-T(H,t)$$
$$\le A^{-1}[\Psi(G,T(H,t)) - \Psi(G,T(G,t))] \quad \text{by (i)}$$

$$= A^{-1}[\Psi(G,T(H,t)) - \Psi(H,T(H,t))] \quad \text{by (6.20)}$$

$$\leq kA^{-1}\|G-H\| \qquad\qquad \text{by (ii)}.$$

Note that conditions (i*) and (ii*) imply that $T: M \times N \longrightarrow \mathbb{R}$ is continuous. To prove Hadamard differentiability we wish to apply Theorem 3.2.4. To do this, consider any compact subset $K \subset V$, sequence $\{H_n\} \subset K$, bounded sequence $\{z_n\} \subset \mathbb{R}$, and sequence $\varepsilon_n \longrightarrow 0$. Then it suffices to prove that

$$(6.21) \qquad \frac{T(G_o + \varepsilon_n H_n, \varepsilon_n z_n) - T(G_o,0)}{\varepsilon_n}$$

is bounded.

Now, for sufficiently large n , $G_o + \varepsilon_n H_n \in M$ and $\varepsilon_n z_n \in N$, so

$$\left| T(G_o + \varepsilon_n H_n, \varepsilon_n z_n) - T(G_o,0) \right|$$

$$\leq \left| T(G_o + \varepsilon_n H_n, \varepsilon_n z_n) - T(G_o, \varepsilon_n z_n) \right|$$

$$+ \left| T(G_o, \varepsilon_n z_n) - T(G_o,0) \right|$$

$$\leq kA^{-1}\|\varepsilon_n H_n\| + A^{-1}|\varepsilon_n z_n|$$

by (i*) and (ii*). Since $\|H_n\|$ and $|z_n|$ are bounded, (6.21) is also bounded. Therefore Theorem 3.2.4 can be applied and it follows that

$$\tau(G) = T(G,0)$$

is Hadamard differentiable at G_o . $\qquad\qquad \Delta$

CHAPTER VII

<u>APPLICATIONS</u>

In this chapter we shall show that certain statistical functionals
are asymptotically normal by applying the techniques developed in the
previous chapters. First, we shall consider M-, L-, and R-estimators.
Besides these, we shall treat a somewhat more complicated statistic, a
gap-compromise estimator.

At times, it is convenient to deal with functionals defined on
$C[0,1]$ rather than on $D[0,1]$. As an example of this we shall consider
sample quantiles, and we shall use an approach which is parallel to and
considerably simpler than that for functionals on $D[0,1]$.

Let us now briefly review the steps that we shall take in the application of von Mises' method to prove the asymptotic normality of a
statistical functional. Suppose that F_n is the empirical d.f. corresponding to a sample with population d.f. F , and consider first the
explicit case:

1. For a statistical functional T , define the induced functional
    $\tau: D[0,1] \longrightarrow \mathbb{R}$ by

$$\tau(G) = T(G \circ F)$$

    for $G \in D[0,1]$.

2. Show that τ is Hadamard differentiable at U , the uniform d.f. in
    $D[0,1]$. This is accomplished by showing that τ can be expressed

as a composition of simpler transformations, each of which is
Hadamard differentiable.

3. Apply Theorem 4.4.2 to conclude that

$$\sqrt{n}(T(F_n)-T(F)) \xrightarrow{D} N(0,\sigma^2)$$

as $n \longrightarrow \infty$, where σ^2 is calculated from the influence curve of
T .

In the implicit case, the procedure is more involved:

1. For a statistical functional T , define the induced functional
$\tau: D[0,1] \longrightarrow \mathbb{R}$ which satisfies

$$\tau(G) = T(G \circ F)$$

for $G \in D[0,1]$. The functional τ is defined implicitly as a
solution $\tau(G) = \theta$ of

$$\Phi(G,\theta) = 0$$

where $\Phi: D[0,1] \times \mathbb{R} \longrightarrow \mathbb{R}$.

2. Show that Φ is Hadamard differentiable at (U,θ_o) where
$\Phi(U,\theta_o) = 0$. This is accomplished by showing that Φ can be expressed as a composition of simpler transformations, each of which
is Hadamard differentiable.

3. Show that Φ satisfies conditions (i) and (ii) of Theorem 6.2.1,
the implicit function theorem for statistical functionals. It follows that τ is Hadamard differentiable at U .

4. As above, apply Theorem 4.4.2 to conclude that as $n \longrightarrow \infty$,

$$\sqrt{n}(T(F_n)-T(F)) \xrightarrow{D} N(0,\sigma^2) .$$

7.1 M-estimators

In section 5.1 we showed that an M-estimator, represented by a statistical functional T , induces a functional $\tau: D[0,1] \longrightarrow \mathbb{R}$ implicitly as a root $\tau(G) = \theta$ of $\Phi(G,\theta) = 0$ where $\Phi: D[0,1] \times \mathbb{R} \longrightarrow \mathbb{R}$ is defined by

$$(7.1) \qquad \Phi(G,\theta) = \int_0^1 \psi(F^{-1}(G^{-1}(x)) - \theta)\,dx \ .$$

Under appropriate conditions on F , and ψ , Theorem 5.1.2 implies that conditions (i) and (ii) of Theorem 6.2.1 are satisfied. Hence, in order to complete the four steps outlined above for implicitly defined statistical functionals, we have only to show that Φ is Hadamard differentiable at (U,θ_o) , where $\Phi(U,\theta_o) = 0$.

<u>Proposition 7.1.1.</u> Let F be a continuous d.f. with piecewise continuous density $F' > 0$. Let ψ be continuous and piecewise differentiable, with bounded derivative ψ' which vanishes outside some bounded interval. Then the function Φ defined by equation (7.1) is Hadamard differentiable at (U,θ_o), where $\Phi(U,\theta_o) = 0$.

<u>Proof</u>: In order to decompose Φ into Hadamard differentiable components, it is first necessary to make a minor adjustment. We wish to apply Proposition 6.1.2 with $L = F^{-1}$, but $L \circ S$ must be defined for S in a neighborhood of U , and $F^{-1} \circ S$ is usually not. However since ψ is constant for large enough and small enough numbers, we can truncate F^{-1} without affecting the value of Φ .

Choose k large enough such that for $|x| \geq k$, $\psi'(x) = 0$. Define the truncated version of F^{-1} by

$$\widetilde{F}^{-1}(x) = \begin{cases} \theta_o - 2k & \text{if } F^{-1}(x) \le \theta_o - 2k \; , \\ F^{-1}(x) & \text{if } |F^{-1}(x) - \theta_o| < 2k \; , \\ \theta_o + 2k & \text{if } F^{-1}(x) \ge \theta_o + 2k \; . \end{cases}$$

We must show that for θ near θ_o ,

$$(7.2) \qquad \psi(\widetilde{F}^{-1}(x) - \theta) = \psi(F^{-1}(x) - \theta)$$

for all $x \in [0,1]$. For $x \in [0,1]$ such that $\widetilde{F}^{-1}(x) = F^{-1}(x)$, (7.2) is trivial. Now suppose that $\widetilde{F}^{-1}(x) \ne F^{-1}(x)$. Then it follows that $|F^{-1}(x) - \theta_o| \ge 2k$ and $|\widetilde{F}^{-1}(x) - \theta_o| = 2k$, so for $|\theta - \theta_o| < k$ we have

$$|F^{-1}(x) - \theta| \ge |F^{-1}(x) - \theta_o| - |\theta_o - \theta|$$

$$\ge 2k - k$$

$$= k \; .$$

Likewise, $|\widetilde{F}^{-1}(x) - \theta| \ge 2k$, and it follows that (7.2) holds for all $x \in [0,1]$ and $|\theta - \theta_o| < k$.

We now have

$$\Phi(G, \theta) = \int_0^1 \psi(\widetilde{F}^{-1}(G^{-1}(x)) - \theta) \, dx$$

for $|\theta - \theta_o| < k$, and we can express Φ as a composition of the following Hadamard differentiable transformations:

$$\gamma_1 : D[0,1] \longrightarrow L^1[0,1] \; , \quad \gamma_1(S) = S^{-1} \; .$$

γ_1 is Hadamard differentiable at $S = U$ by Proposition 6.1.1.

$$\gamma_2 : L^1[0,1] \longrightarrow L^1[0,1] \; , \quad \gamma_2(S) = F^{-1}(S) \; .$$

γ_2 is Hadamard differentiable at $S = U$ by Proposition 6.1.2.

$$\gamma_3: \ L^1[0,1] \times \mathbb{R} \longrightarrow L^1[0,1] \ , \qquad \gamma_3(S,\theta) = S-\theta.$$

γ_3 is Fréchet differentiable since it is linear and continuous.

$$\gamma_4: \ L^1[0,1] \longrightarrow L^1[0,1] \ , \qquad \gamma_4(S) = \psi \circ S \ .$$

γ_4 is Hadamard differentiable at $S = \tilde{F}^{-1} - \theta_0$ by Proposition 6.1.2.

$$\gamma_5: \ L^1[0,1] \longrightarrow \mathbb{R} \ , \qquad \gamma_5(S) = \int_0^1 S(x)\,dx \ .$$

γ_5 is Fréchet differentiable since it is linear and continuous.

We can write

$$\Phi(G,\theta) = \gamma_5(\gamma_4(\gamma_3(\gamma_2(\gamma_1(G)),\theta)))\ ,$$

so, by applying the chain rule, it follows that Φ is Hadamard differentiable at (U,θ_0) .

$$\Delta$$

Theorem 7.1.2. Let F be a continuous d.f. with piecewise continuous density $F' > 0$. Let ψ be nondecreasing, continuous, and piecewise differentiable, with bounded derivative ψ' such that $0 < m \leq \psi'(x)$, m constant, for x in some neighborhood of 0 and $\psi'(x) = 0$ outside some bounded interval. If F_n is the empirical d.f. corresponding to a sample $X_1,\ldots,X_n$ with d.f. F , then the M-estimator $T(F_n) = \theta$ defined as a root of

$$\int_0^1 \psi(F_n^{-1}(x)-\theta)\,dx = 0$$

satisfies

$$\sqrt{n}(T(F_n)-T(F)) \xrightarrow{\ D\ } N(0,\sigma^2)$$

as $n \longrightarrow \infty$, where $\sigma^2 = \text{Var } IC(X_1;F,T)$. The influence curve IC is given by

$$IC(x;F,T) = \frac{\psi(x-T(F))}{\int \psi'(x-T(F))\,dF(x)} \quad .$$

Proof: For the computation of the influence curve see Huber (1981). The conclusion of the theorem then follows by applying Proposition 7.1.1, Theorem 5.1.2, Theorem 6.2.1, and Theorem 4.4.2, in the manner outlined above. Δ

Example 7.1.3. An M-estimator of location proposed by Huber has

$$\psi(x) = \begin{cases} -c & \text{if } \ x < -c \\[2ex] x & \text{if } \ -c \le x \le c \\[2ex] c & \text{if } \ c < x \end{cases}$$

where c is a positive constant. This estimator is asymptotically normal for those population d.f.'s F which satisfy the hypotheses of Theorem 7.1.2. Δ

7.2 L-estimators

In section 5.2 we defined an L-estimator as an explicitly defined statistical functional which induces a functional $\tau: D[0,1] \longrightarrow \mathbb{R}$ of the form

$$(7.3) \qquad \tau(G) = \int_0^1 h(F^{-1}(G^{-1}(x)))\ m(x)\,dx \ ,$$

$G \in D[0,1]$, where F is a population d.f., h is a real valued

function, and m is the density of a (signed) measure M on [0,1] .
Here we shall show that under appropriate conditions, τ is Hadamard
differentiable at the uniform d.f. U .

<u>Proposition 7.2.1</u>. Let $h: \mathbb{R} \longrightarrow \mathbb{R}$ be continuous and piecewise diff-
erentiable with bounded derivative. Let $m \in L^2[0,1]$ have support in
$[\alpha,1-\alpha]$ for some $\alpha > 0$, and suppose that F is an absolutely continuous increasing d.f.. Then the functional $\tau: [0,1] \longrightarrow \mathbb{R}$ defined
by (7.3) is Hadamard differentiable at the uniform d.f. $U \in D[0,1]$.

<u>Proof</u>: To be able to show that τ is a composition of Hadamard differ-
entiable functions, we must make an adjustment as in Proposition 7.1.1.
Here we wish to apply Proposition 6.1.2 with $L = h \circ F^{-1}$ at $S = U$,
so $L \circ S$ must be defined for S in a neighborhood of U, and $h \circ F^{-1} \circ S$
is usually not. However the fact that $m(x) = 0$ for x outside
$[\alpha,1-\alpha]$ means that $\tau(G)$ is not affected by the behavior of $h \circ G^{-1}(x)$
for x near 0 or 1 . Now if $\|G-U\| < \alpha/2$, then $\|G^{-1}-U\| < \alpha/2$,
and if we define

$$\tilde{h}(y) = \begin{cases} h(F^{-1}(-\frac{\alpha}{2})) & \text{if } y < F^{-1}(-\frac{\alpha}{2}) \\ \\ h(y) & \text{if } F^{-1}(-\frac{\alpha}{2}) \leq y \leq F^{-1}(\frac{\alpha}{2}) \\ \\ h(F^{-1}(\frac{\alpha}{2})) & \text{if } y > F^{-1}(\frac{\alpha}{2}) \end{cases}$$

and

$$\tilde{\tau}(G) = \int_0^1 \tilde{h}(F^{-1}(G^{-1}(x)))\, m(x)dx ,$$

then $\tilde{\tau}(G) = \tau(G)$ for $\|G-U\| < \alpha/2$. Therefore it suffices to show that
$\tilde{\tau}$ is Hadamard differentiable at U .

   We shall now express $\tilde{\tau}$ as a composition of the following Hadamard
differentiable transformations:

$$\gamma_1: D[0,1] \longrightarrow L^2[0,1] \ , \qquad \gamma_1(S) = S^{-1} \ .$$

γ_1 is Hadamard differentiable at U by Proposition 6.1.1.

$$\gamma_2: L^2[0,1] \longrightarrow L^2[0,1] \ , \qquad \gamma_2(S) = \tilde{h} \circ F^{-1} \circ S \ .$$

γ_2 is Hadamard differentiable at U by Proposition 6.1.2.

$$\gamma_3: L^2[0,1] \longrightarrow \mathbb{R} \ , \quad \gamma_3(S) = \int_0^1 S(x) m(x) dx \ .$$

γ_3 is linear and continuous, thus Fréchet differentiable.

We can express $\tilde{\tau}(G)$ as

$$\tilde{\tau}(G) = \gamma_3(\gamma_2(\gamma_1(G))) \ ,$$

so, by the chain rule, $\tilde{\tau}$ is Hadamard differentiable at U , and since τ agrees with $\tilde{\tau}$ in a neighborhood of U , τ is also Hadamard differentiable at U . $\hfill \Delta$

<u>Theorem 7.2.2</u>. Let T be the L-estimator defined by the statistical functional

$$T(F) = \int_0^1 h(F^{-1}(x)) m(x) dx$$

where $h: \mathbb{R} \longrightarrow \mathbb{R}$ is continuous and piecewise differentiable with bounded derivative, and $m \in L^2[0,1]$ has support in $[\alpha, 1-\alpha]$ for some $\alpha > 0$. Let F be an absolutely continuous, increasing d.f.. If $X_1, \ldots, X_n$ is a sample from F , and F_n is the corresponding empirical d.f., then

$$\sqrt{n}(T(F_n) - T(F)) \xrightarrow{\ \mathcal{D}\ } N(0, \sigma^2)$$

with

$$\sigma^2 = \operatorname{Var} IC(X_1; F, T)$$

where

$$IC(x;F,T) = \int_{-\infty}^{x} h'(y)m(F(y))dy - \int_{-\infty}^{\infty} (1-F(y))h'(y)m(F(y))dy \ .$$

<u>Proof</u>: The computation of the influence curve $IC(x;F,T)$ can be found in Huber (1981). The proof of the theorem then follows by applying Proposition 7.2.1 and Theorem 4.4.2. $\qquad\qquad \Delta$

<u>Example 7.2.3</u>. The α-trimmed mean for $0 < \alpha < \frac{1}{2}$ is an L-estimator with $h(x) = x$ and $m(x) = 1/(1-2\alpha)$ for $x \in (\alpha, 1-\alpha)$ and $m(x) = 0$ elsewhere. This estimator is asymptotically normal for those d.f.'s which satisfy the hypotheses of Theorem 7.2.2. $\qquad\qquad \Delta$

7.3 R-estimators

An R-estimator, as we saw in section 5.3, is represented by a statistical functional T defined implicitly as a root $T(F) = \theta$ of

$$(7.4) \qquad \int_{0}^{1} J(x - F(2\theta - F^{-1}(x)))dx \ .$$

For a given continuous, increasing d.f. F , this induces a functional τ on $D[0,1]$ defined as a root $\tau(G) = \theta$, $G \in D[0,1]$, of $\Phi(G,\theta) = 0$ where

$$(7.5) \qquad \Phi(G,\theta) = \int_{0}^{1} J[x-G(F(2\theta - F^{-1}(G^{-1}(x))))]dx \ .$$

The equation $\Phi(G,\theta) = 0$ may have multiple roots, so in section 5.4 we introduced modified versions

$$\hat{\Phi}(G,\theta) = \int_{0}^{1} J[x-\hat{G}(F(2\theta - F^{-1}(G^{-1}(x))))]dx$$

and

$$\check{\Phi}(G,\theta) = \int_{0}^{1} J[x-\check{G}(F(2\theta - F^{-1}(G^{-1}(x))))]dx$$

with corresponding uniquely defined functionals $\hat{\tau}$ and $\check{\tau}$ satisfying respectively

$$\hat{\Phi}(G,\hat{\tau}(G)) = 0$$

and

$$\check{\Phi}(G,\check{\tau}(G)) = 0 \ ,$$

for $G \in D[0,1]$. The functionals $\hat{\tau}$ and $\check{\tau}$ satisfy

$$\hat{\tau}(G) \leq \tau(G) \leq \check{\tau}(G)$$

for all $G \in D[0,1]$, and

$$\check{\tau}(U) = \tau(U) = \hat{\tau}(U)$$

where $U(x) = x$.

We shall first prove the Hadamard differentiability of $\hat{\Phi}$ and $\check{\Phi}$. Throughout this section we shall assume that the score function J has been extended to be an odd function defined on $\mathbb{R}$.

Proposition 7.3.1. Suppose that J is odd, continuous, and piecewise differentiable on $\mathbb{R}$, with bounded, piecewise continuous derivative J'. Let F be a d.f. on $\mathbb{R}$ for which $\Gamma(x,\theta) = F(2\theta-F^{-1}(x))$ satisfies the hypotheses of Proposition 6.1.3 for θ_o , where $\hat{\Phi}(U,\theta_o) = \check{\Phi}(U,\theta_o) = 0$. Then $\hat{\Phi}$ and $\check{\Phi}$ are Hadamard differentiable at (U,θ_o) .

Proof: Consider the function $\hat{\Phi}$. As usual we shall express this function as a composition of Hadamard differentiable transformations.

$$\gamma_1: D[0,1] \longrightarrow L^2[0,1] \ , \quad \gamma_1(G) = G^{-1} \ .$$

γ_1 is Hadamard differentiable at $G = U$ by Proposition 6.1.1.

$$\gamma_2: L^2[0,1] \times \mathbb{R} \longrightarrow L^1[0,1] \ , \quad \gamma_2(G,\theta) = F(2\theta-F^{-1}(G)) \ .$$

γ_2 is Fréchet differentiable at (U,θ_o) by Proposition 6.1.3.

$$\hat{\gamma}_3: D[0,1] \times L^1[0,1] \longrightarrow L^1[0,1] \ , \quad \hat{\gamma}_3(G,\theta) = \hat{G} \circ Q \ .$$

$\hat{\gamma}_3$ is Hadamard differentiable at (U,Q) where $Q = F(2\theta_o - F^{-1}(\cdot))$ by Proposition 6.1.7.

$$\gamma_4: L^1[0,1] \longrightarrow L^1[0,1] \ , \quad \gamma_4(G) = U - G \ .$$

γ_4 is linear and continuous, and hence Fréchet differentiable.

$$\gamma_5: L^1[0,1] \longrightarrow L^1[0,1] \ , \quad \gamma_5(G) = J \circ G \ .$$

γ_5 is Hadamard differentiable by Proposition 6.1.2.

$$\gamma_6: L^1[0,1] \longrightarrow \mathbb{R} \ , \quad \gamma_6(G) = \int_0^1 G(x)\,dx \ .$$

γ_6 is linear and continuous and hence Fréchet differentiable.

We can now write

$$(7.6) \qquad \hat{\Phi}(G,\theta) = \gamma_6 \circ \gamma_5 \circ \gamma_4 \circ \hat{\gamma}_3(G, \gamma_2(\gamma_1(G),\theta)) \ ,$$

so it follows by Proposition 3.1.2 (chain rule) that $\hat{\Phi}$ is Hadamard differentiable at (U,θ_o) .

For $\check{\Phi}$, the only component which will differ from (7.6) is

$$\check{\gamma}_3(G,Q) = \check{G} \circ Q \ ,$$

which will replace $\hat{\gamma}_3$. By Proposition 6.1.7, the derivatives of $\hat{\gamma}_3$ and $\check{\gamma}_3$ are equal at (U,Q) , where $Q = F(2\theta_o - F^{-1}(\cdot))$. Hence $\check{\Phi}$ is Hadamard differentiable at (U,θ_o) and $\hat{\Phi}'_{(U,\theta_o)} = \check{\Phi}'_{(U,\theta_o)}$. $\qquad \Delta$

The proof of Proposition 7.3.1 can easily be modified to accommodate distributions that are concentrated on a finite interval $[a,b]$. In this case we shall define

$$F^{-1}(y) = \sup\{a, \inf\{b, x: F(x) \geq y\}\} .$$

<u>Proposition 7.3.2</u>. Let F be a symmetric d.f. concentrated on $[a,b]$ with continuous density bounded away from zero and infinity on $[a,b]$. Let $\hat{\Phi}$ and $\check{\Phi}$ be as above with J odd, continuous, and piecewise differentiable on $\mathbb{R}$, with bounded, piecewise continuous derivative J'. Then $\hat{\Phi}$ and $\check{\Phi}$ are Hadamard differentiable at (U,θ_o), where $\theta_o = \frac{a+b}{2}$ is the center of symmetry of F. The derivatives of $\hat{\Phi}$ and $\check{\Phi}$ are equal at (U,θ_o).

<u>Proof</u>: The proof follows along the lines of that of Proposition 7.3.1 with a few minor changes. Consider

$$G(F(2\theta - F^{-1}(G^{-1}(x)))) .$$

We can decompose this as follows:

$$\gamma_1 : D[0,1] \longrightarrow L^1[0,1] , \qquad \gamma_1(G) = G^{-1} .$$

γ_1 is Hadamard differentiable at U by Proposition 6.1.1.

$$\gamma_2 : L^1[0,1] \longrightarrow L^1[0,1] , \qquad \gamma_2(G) = F^{-1} \circ G .$$

γ_2 is Hadamard differentiable at $G = U$ by Proposition 6.1.2.

$$\gamma_3 : L^1[0,1] \longrightarrow L^1[0,1] , \qquad \gamma_3(G) = F \circ G .$$

γ_3 is Hadamard differentiable at $G = 2\theta - F^{-1}$ by Proposition 6.1.2.

$$\gamma_4 : D[0,1] \times L^1[0,1] \longrightarrow L^1[0,1] , \quad \gamma_4(G,Q) = G \circ Q .$$

γ_4 is Hadamard differentiable at (U,Q), where

99

$$Q(x) = F(2\theta_o - F^{-1}(x)) \ ,$$

because $Q' = -1$ and we can apply Proposition 6.1.6.

The rest follows as in Proposition 7.3.1. $\qquad\qquad \Delta$

We shall use the implicit function theorem to show that $\hat{\tau}$ and $\check{\tau}$ are Hadamard differentiable, which will then imply that τ is Hadamard differentiable by an application of the following lemma.

<u>Lemma 7.3.4.</u> Let V be a topological vector space and suppose that $\tau, \hat{\tau}, \check{\tau}: V \longrightarrow \mathbb{R}$ are defined in a neighborhood N of $G_o \in V$. Suppose that for $G \in N$,

$$\hat{\tau}(G) \leq \tau(G) \leq \check{\tau}(G)$$

with equality for $G = G_o$. If $\hat{\tau}$ and $\check{\tau}$ are S-differentiable at G_o with $\hat{\tau}'_{G_o} = \check{\tau}'_{G_o}$, then τ is S-differentiable at G_o with

$$\tau'_{G_o} = \hat{\tau}'_{G_o} = \check{\tau}'_{G_o} \ .$$

<u>Proof</u>: Let $K \subset V$ be in S . Then for $H \in K$ and $t \in \mathbb{R}$ let

$$\text{Rem}(tH) = \tau(G_o + tH) - \tau(G_o) - t\tau'_{G_o}(H)$$

and define $\hat{\text{Rem}}(tH)$ and $\check{\text{Rem}}(tH)$ in a similar manner with $\hat{\tau}$ and $\check{\tau}$, respectively, replacing τ . Then for $t \neq 0$,

$$\frac{\hat{\text{Rem}}(tH)}{t} \leq \frac{\text{Rem}(tH)}{t} \leq \frac{\check{\text{Rem}}(tH)}{t} \ .$$

Since both $\dfrac{\hat{\text{Rem}}(tH)}{t}$ and $\dfrac{\check{\text{Rem}}(tH)}{t}$ tend to zero uniformly for $H \in K$ as $t \longrightarrow 0$, it follows that $\dfrac{\text{Rem}(tH)}{t}$ does likewise. Therefore τ is S-differentiable at G_o with $\tau'_{G_o} = \hat{\tau}'_{G_o} = \check{\tau}'_{G_o}$. $\qquad \Delta$

<u>Theorem 7.3.5</u>. Suppose that J is continuous, odd, increasing, and piecewise differentiable on $\mathbb{R}$, with bounded, piecewise continuous derivative J' such that there exists $m > 0$ with $J'(x) \geq m$ for x in some neighborhood of zero. Let F be a d.f. on $\mathbb{R}$ for which $\Gamma(x,\theta) = F(2\theta - F^{-1}(x))$ satisfies the hypotheses of Proposition 6.1.3 for θ_o , where $\Phi(U,\theta_o) = 0$. If F_n is the empirical d.f. corresponding to a sample $X_1,\ldots,X_n$ with d.f. F , then the R-estimator $T(F_n) = \theta$ defined as a root of

$$\int_0^1 J(x - F_n(2\theta - F_n^{-1}(x)))dx = 0$$

satisfies

$$(7.7) \qquad\qquad \sqrt{n}(T(F_n) - T(F)) \xrightarrow{\;D\;} N(0,\sigma^2)$$

where $\sigma^2 = \operatorname{Var} IC(X_1;F,T)$.

<u>Proof</u>: By Proposition 7.3.1, Theorem 5.4.4, and Theorem 6.2.1, it follows that $\hat{\tau}$ and $\check{\tau}$ are Hadamard differentiable at U with

$$\hat{\tau}'_U = \check{\tau}'_U \ .$$

Hence by Lemma 7.3.4, τ is Hadamard differentiable at U and we can apply Theorem 4.4.2, which implies (7.7). $\qquad\qquad\qquad\qquad\qquad \Delta$

The computation of the influence curve in Theorem 7.3.3 can be found in Fernholz (1979) or Huber (1981). The equation for $IC(x;F,T)$ is somewhat cumbersome for general R-estimators, and we shall omit it here. However for symmetric d.f. F , for which R-estimators are most commonly used, the equation for the influence curve is more tractable and we have

$$(7.8) \qquad\qquad IC(x;F,T) = \frac{J(2F(x)-1)}{2 \int J'(2F(x)-1)f^2(x)dx}$$

where $f = F'$.

<u>Corollary 7.3.6</u>. Let J and T be as in Theorem 7.3.5 and suppose that F is symmetric about θ_o with bounded, regular density $F' > 0$. Then

$$\sqrt{n}(T(F_n)-T(F)) \xrightarrow{\ \mathcal{D}\ } N(0,\sigma^2)$$

where $\sigma^2 = \text{Var } IC(X_1;F,T)$.

<u>Proof</u>: If F is symmetric about θ_o then $\Gamma(x,\theta) = F(2\theta-F^{-1}(x))$ satisfies the hypotheses of Proposition 6.1.3 for θ_o . The proof then follows directly from Theorem 7.3.5. Δ

   Another simplified case is for a d.f. F which is concentrated on a bounded interval.

<u>Corollary 7.3.7</u>. Let J and T be as in Theorem 7.3.5 and suppose that F is symmetric and concentrated on an interval [a,b] with positive, bounded density F' . Then

$$\sqrt{n}(T(F_n)-T(F)) \xrightarrow{\ \mathcal{D}\ } N(0,\sigma^2)$$

where $\sigma^2 = \text{Var } IC(X_1;F,T)$.

<u>Proof</u>: Follows directly from Proposition 7.3.2, Lemma 7.3.4, and Theorem 4.4.2. Δ

7.4 Functionals on C[0,1]: sample quantiles

   Sample quantiles can be considered to be either M-estimators or L-estimators, but in neither case do they satisfy the conditions necessary for the asymptotic normality results which were proved in section 7.1 and 7.2. To handle sample quantiles, we shall show that they induce Hadamard differentiable functionals on C[0,1] and we shall treat these functionals in a manner which is parallel to and simpler than the treatment of functionals on D[0,1] .

For a d.f. F and $0 < q < 1$, let the q-th quantile of F be defined by the functional

$$T(F) = F^{-1}(q) \ .$$

For our purposes, it is convenient to express $T(F)$ implicitly as a solution $T(F) = \theta$ of $F(\theta)-q = 0$, where in case of discontinuities of F , θ is taken to be the point at which the left hand side of the equation changes sign.

For a given continuous d.f. F , T induces a functional τ on $C[0,1]$ impliclty as a root $\tau(G) = \theta$ of

$$\Phi(G,\theta) = 0 \ ,$$

where $\Phi: C[0,1] \times \mathbb{R} \longrightarrow \mathbb{R}$ is defined by

$$\Phi(G,\theta) = G(F(\theta))-q \ .$$

We saw in Chapter IV that for a given continuous d.f. F , the empirical d.f. F_n corresponding to a sample $X_1,\ldots,X_n$ can be expressed as $F_n = U_n \circ F$, where U_n is the empirical d.f. of $Y_1,\ldots,Y_n$ with $Y_i = F(X_i)$, $i = 1,\ldots,n$. Since U_n is not continuous, we must modify it before we can apply the functional τ . Let $[\cdot]$ represent the greatest-integer-less-than function (so for any integer n , $[n] = n-1$), let $0 < q < 1$, and define $\nu_n = q-[nq]/n$. Let $Y_{(1)} \leq Y_{(2)} \leq \cdots \leq Y_{(n)}$ be the ordered Y_i's , and define the modified version $\tilde{U}_n$ of U_n to be the d.f. on $[0,1]$ corresponding to a uniform distribution of mass ν_n on $[0,Y_{(1)}]$, a uniform distribution of mass $\frac{1}{n}$ on $[Y_{(i)},Y_{(i+1)}]$, $i = 1,\ldots,n-1$, and a uniform distribution of mass $\frac{1}{n} - \nu_n$ on $[Y_{(n)},1]$. If F is continuous, then $\tilde{U}_n$ will almost surely be continuous. It can easily be checked that

$$\|\tilde{U}_n - U_n^*\| \leq \frac{1}{n} \ ,$$

where U_n^* is the continuous version of U_n that we used in Chapter IV, and this inequality implies that all the results of that chapter for U_n^* are also valid for $\tilde{U}_n$.

If F is increasing near $F^{-1}(q)$, then

$$\tau(\tilde{U}_n) = T(F_n) ,$$

so the results of Chapter IV imply that in order to prove the asymptotic normality of $\sqrt{n}(T(F_n)-T(F))$, it suffices to show that τ is Hadamard differentiable at $U \in C[0,1]$. We shall accomplish this by first showing that ϕ is Hadamard differentiable at (U,θ_o) , where $\phi(U,\theta_o) = 0$, and then applying Theorem 6.2.1.

Proposition 7.4.1. If F is differentiable at θ_o , then ϕ is Hadamard differentiable at (U,θ_o) with Taylor expansion

$$\phi(U+tH,\theta_o+th) = \phi(U,\theta_o) + tH(F(\theta_o)) = thF'(\theta_o) + \text{Rem}(tH,th) .$$

Proof: Let $K \subset C[0,1]$ be compact and let $H \in K$. Let $h \in \mathbb{R}$ with $|h| \leq k$, k constant. Then

$$\frac{\text{Rem}(tH,th)}{t} = \frac{\phi(U+tH,\theta_o+th)-\phi(U,\theta_o)-tH(F(\theta_o))-thF'(\theta_o)}{t}$$

$$= \frac{F(\theta_o+th)-F(\theta_o)}{t} - hF'(\theta_o) + H(F(\theta_o+th)) - H(F(\theta_o)) .$$

Since F is differentiable and therefore continuous at θ_o , and since K is equicontinuous, this last expression tends to zero uniformly for $H \in K$ and $|h| < k$ as $t \longrightarrow 0$. $\qquad\qquad \Delta$

The next step is to show that conditions (i) and (ii) of Theorem 6.2.1 hold. Unfortunately, if G is not strictly monotone (i) will fail because

$$\Phi(G,\theta) = 0$$

may have multiple roots. Therefore we shall define $\hat{\Phi}$ and $\check{\Phi}$ as was done for R-estimators. Let

$$\hat{\Phi}(G,\theta) = \hat{G}(F(\theta))-q$$

(7.9)

$$\check{\Phi}(G,\theta) = \check{G}(F(\theta))-q$$

where $\hat{G}$ and $\check{G}$ are the transformations defined in Chapter V (Definition 5.4.1). Then

$$\hat{\Phi}(G,\theta) \geq \Phi(G,\theta) \geq \check{\Phi}(G,\theta)$$

with equality for $G = U$.

In order to prove that $\hat{\Phi}$ and $\check{\Phi}$ are Hadamard differentiable at U with the same derivative as Φ , it suffices to prove

Lemma 7.4.2. Let $K \subset C[0,1]$ be compact and $k \in \mathbb{R}$ be positive. Then, if F is continuous at θ_o and $\Phi(U,\theta_o) = 0$,

$$\frac{\hat{\Phi}(U+tH,\theta_o+th) - \Phi(U+tH,\theta_o+th)}{t} \longrightarrow 0$$

as $t \longrightarrow 0$, uniformly for $H \in K$ and $|h| \leq k$.

A similar result holds for $\check{\Phi}$.

Proof: We have

$$0 \leq \hat{\Phi}(U+tH,\theta_o+th) - \Phi(U+tH,\theta_o+th)$$

$$= (U+tH)\hat{}(F(\theta_o+th)) - (U+tH)(F(\theta_o+th)) \ .$$

Let $z = F(\theta_o+th)$, then as in the proof of Proposition 6.1.7 there exists a z' such that $|z-z'| \leq 2tk/(1-\alpha)$, where $k = \sup \{ \|H\| : H \in K\}$, and

$$(U+tH)\hat{\ }(z) - (U+tH)(z) \leq t[H(z')-H(z)] + t^2 .$$

Since F is continuous at θ_o , both z and z' tend to $F(\theta_o)$ as $t \longrightarrow 0$. Since K is equicontinuous $H(z)-H(z') \longrightarrow 0$ uniformly for $H \in K$ and $|h| < k$ as $t \longrightarrow 0$. Hence the conclusion follows.

Similarly for $\check{\phi}$. Δ

Now we shall show that $\hat{\phi}$ and $\check{\phi}$ satisfy the hypotheses of Theorem 6.2.1, the implicit function theorem.

<u>Proposition 7.4.3.</u> Let N be a neighborhood of θ_o in $\mathbb{R}$ and assume that there exist positive constants a and b such that for $x,y \in N$, $x \leq y$,

$$a(y-x) \leq F(y)-F(x) \leq b(y-x) .$$

Then $\hat{\phi}$ and $\check{\phi}$, defined in (7.9), satisfy conditions (i) and (ii) of Theorem 6.2.1.

<u>Proof</u>: For (i), we have that for $\sigma < \theta$

$$\hat{\phi}(G,\theta) - \hat{\phi}(G,\sigma) = \hat{G}(F(\theta)) - \hat{G}(F(\sigma))$$

and

$$\alpha[F(\theta)-F(\sigma)] \leq \hat{G}(F(\theta))-\hat{G}(F(\sigma)) \leq \alpha^{-1}[F(\theta)-F(\sigma)]$$

so

$$\alpha a(\theta-\sigma) \leq \hat{G}(F(\theta))-\hat{G}(F(\sigma)) \leq \alpha^{-1}b(\theta-\sigma) .$$

(ii) follows trivially.

Analogously for $\check{\phi}$. Δ

Now we let $\hat{\tau}(G)$ and $\check{\tau}(G)$ be the roots of $\hat{\phi}(G,\theta) = 0$ and $\check{\phi}(G,\theta) = 0$ respectively. Then we have

$$\hat{\tau}(G) \leq \tau(G) \leq \check{\tau}(G)$$

with equality for $G = U$.

<u>Proposition 7.4.4.</u> Let F be a d.f. which is differentiable at θ_o , where $\Phi(U,\theta_o) = 0$, and suppose that there exists a neighborhood N of θ_o such that if $x,y \in N$ and $x \leq y$ then

$$a(y-x) \leq F(y)-F(x) \leq b(y-x)$$

for some positive constants a,b . Then the functional τ satisfying $\Phi(G,\tau(G)) = 0$, is Hadamard differentiable at U .

<u>Proof</u>: By Lemma 7.4.2, $\hat{\Phi}$ and $\check{\Phi}$ are Hadamard differentiable at U and their derivatives coincide there. Hence Proposition 7.4.3 can be applied to $\hat{\Phi}$ and $\check{\Phi}$, so by the implicit function theorem (Theorem 6.2.1), $\hat{\tau}$ and $\check{\tau}$ are Hadamard differentiable at U with $\hat{\tau}'_U = \check{\tau}'_U$. Now apply Lemma 7.3.4 and the conclusion follows. Δ

We have now proved that τ is Hadamard differentiable at U . Recall that in Example 2.3.2 we showed that this functional is not Fréchet differentiable.

<u>Theorem 7.4.5.</u> Let $0 < q < 1$, let F be a d.f. which is differentiable at $\theta_o = F^{-1}(q)$, and suppose that there exists a neighborhood N of θ_o such that for $x,y \in N$, $x \leq y$,

$$a(y-x) \leq F(y)-F(x) \leq b(y-x)$$

for some positive constants a,b . If $X_1,\ldots,X_n$ is a sample from F and F_n is the emprical d.f., then the q-th sample quantile $T(F_n) = F_n^{-1}(q)$ satisfies

$$\sqrt{n}(T(F_n)-T(F)) \xrightarrow{\mathcal{D}} N(0,\sigma^2)$$

as $n \longrightarrow \infty$, where

$$\sigma^2 = \text{Var } IC(X_1;T,F)$$

with

$$IC(x;T,F) = \begin{cases} (q-1)/F'(F^{-1}(q)) & \text{if } x < F^{-1}(q) \\ \\ q/F'(F^{-1}(q)) & \text{if } x > F^{-1}(q) \end{cases} .$$

Proof: For the calculation of the influence curve see Huber (1981). The proof then follows directly from Proposition 7.4.4 and Corollary 4.4.3. Δ

7.5 Truncated d.f.'s and modified estimators

Much of the difficulty in showing that a particular statistical functional is Hadamard differentiable or in showing that it satisfies the hypotheses of the implicit function theorem is caused by the behavior of the functions involved at $\pm \infty$. Accordingly, it is often possible to greatly simplify proofs by truncating the population d.f. or by modifying the estimator in some way. Since the values at which the truncation occurs can be arbitrarily large, this procedure need have no practical effect on the estimator. This simplifying technique permits the application of our methodology to functionals which might otherwise be analytically intractable, and certainly the results are as meaningful as the results of the simulations which are often used in such cases as a last resort.

We shall present here two examples of modifications.

Example 7.5.1. The normal scores rank estimator is an R-estimator defined by equation(7.4), with $J = N^{-1}$ where N is the d.f. for the standard normal distribution. Since $J(x)$ is not defined for $x = 0,1$, the hypotheses of Theorem 7.3.5 are not satisfied. Now let us modify J

at some large number α , by defining a new function $\tilde{J}$ to be odd and continuous with $\tilde{J}(x) = J(x)$ for $|J(x)| \leq \alpha$, and with derivative $\tilde{J}'(x) = 1$ elsewhere. Then if we replace J by $\tilde{J}$, the conditions of Theorem 7.3.5 are satisfied, and consequently the modified estimator is asymptotically normal.

If α is chosen to be large enough, the modified estimator and the original normal scores estimator will have the same value for samples of any reasonable size. $\qquad\qquad\Delta$

<u>Example 7.5.2</u>. <u>Gap-compromise estimators</u> are location estimators that were developed by R. Guarino (1980) and are constructed to be optimal, in some sense, when the population d.f. can be one of two possible choices. We include these estimators here because their mathematical complexity provides a good test of our methodology. The gap-compromise estimator we shall consider is called a G-estimator.

Let F_1 and F_2 be d.f.'s with densities f_1 and f_2 respectively, and let $0 < \lambda < 1$. To save one step, we shall define the G-estimator in terms of the functional which it induces on $D[0,1]$. Let F be a d.f. and define $\Phi: D[0,1] \times \mathbb{R} \longrightarrow \mathbb{R}$ by

$$(7.10) \quad \Phi(G,\theta) = -\int_{-m}^{m} \frac{\lambda A(G(F(z)))M_1(G,\theta)(z) + (1-\lambda)M_2(G,\theta)(z)}{\lambda A(G(F(z))) + 1-\lambda} \, dz \, ,$$

where $m = \infty$,

$$A(x) = \left[\frac{f_2(F_2^{-1}(x))}{f_1(F_1^{-1}(x))}\right]^2 \, ,$$

and

$$M_i(G,\theta)(z) = \frac{\displaystyle\int_0^{G(F(z))} w_i(F^{-1}(G^{-1}(x))-\theta)\,dx}{\displaystyle\int_0^1 w_i(F^{-1}(G^{-1}(x))-\theta)\,dx}$$

with

$$w_i(x) = \frac{-f_i'(x)}{x\,f_i(x)} \; ,$$

for $i = 1,2$. The G-estimator $T(F)$ is defined implicitly by $T(F) = \theta_o$ where

$$\Phi(U,\theta_o) = 0$$

for U the uniform d.f. on $[0,1]$. We can define the induced functional τ on $D[0,1]$ as a root $\tau(G) = \theta$ of

$$\Phi(G,\theta) = 0 \; .$$

To prove that $\sqrt{n}(T(F_n)-T(F))$ is asymptotically normal, it suffices to show that τ is Hadamard differentiable at U.

The first simplification that we shall undertake is to let m in (7.10) be a finite number and to consider the d.f.'s F, F_1, and F_2 to be concentrated on $[-m,m]$. We shall first prove that Φ is Hadamard differentiable at (U,θ_o).

Consider the following transformations:

$$\gamma_1 : D[0,1] \longrightarrow D[0,1] \; , \qquad \gamma_1(G) = G \circ F \; .$$

γ_1 is linear and continuous, and hence Hadamard differentiable.

$$\gamma_2 : D[0,1] \longrightarrow L^q[0,1] \; , \qquad \gamma_2(G) = A \circ G \; .$$

Assuming sufficient regularity conditions of A, γ_2 is Hadamard differentiable at $G = F$ by Proposition 6.1.2. We have now shown that $G \longrightarrow A \circ G \circ F$ is Hadamard differentiable at $G = U$. The denominator of the integral in (7.10) is

$$\lambda \gamma_2 \circ \gamma_1 (G)(z) + 1 - \lambda \; .$$

$$\gamma_3 : L^q[0,1] \longrightarrow L^\infty[0,1] \; , \qquad \gamma_3(S) = \frac{1}{S} \; .$$

110

If $S = \lambda A \circ G \circ F + (1-\lambda) \geq a > 0$, then γ_3 is Hadamard differentiable at S by Proposition 6.1.2. We have now proved that the transformation from $D[0,1]$ to $L^\infty[0,1]$ defined by

$$G \longrightarrow \frac{1}{\lambda A \circ G \circ F + (1-\lambda)}$$

is Hadamard differentiable at $G = U$.

Now consider the functions $M_i : D[0,1] \times \mathbb{R} \longrightarrow L^P[0,1]$, $i = 1,2$, and the following transformations:

$$\gamma_4 : D[0,1] \times \mathbb{R} \longrightarrow \mathbb{R} \ , \quad \gamma_4(G,\theta) = \int_0^1 w_i(F^{-1}(G^{-1}(x)) - \theta)dx \ .$$

γ_4 is Hadamard differentiable at (U,θ_o) since it has the form of an M-estimator.

$$\gamma_5 : D[0,1] \times \mathbb{R} \longrightarrow \mathbb{R} \ , \qquad \gamma_5(G,\theta) = w_i(F^{-1} \circ G^{-1} - \theta) \ .$$

γ_5 is Hadamard differentiable at (U,θ_o) for the same reason as γ_4 .

$$\gamma_6 : L^P[0,1] \longrightarrow C[0,1] \ , \quad \gamma_6(S) = \int_0^{\cdot} S(x)dx \ ,$$

where $\cdot$ serves as a place holder for the argument of the function. γ_6 is linear and continuous, hence Hadamard differentiable.

$$\gamma_7 : C[0,1] \times L^P[0,1] \longrightarrow L^P[0,1] \ , \quad \gamma_7(G,H) = G \circ H \ .$$

γ_7 is Hadamard differentiable by Proposition 6.1.3. We now have,

$$(7.11) \qquad M_i(G,\theta) = \frac{\gamma_7(\gamma_6 \circ \gamma_5(G,\theta), \gamma_1(G))}{\gamma_4(G,\theta)} \ ,$$

so $M_i : D[0,1] \times \mathbb{R} \longrightarrow L^P[0,1]$ is Hadamard differentiable at (U,θ_o) , for $i = 1,2$.

$$\gamma_8 : L^q[0,1] \times L^P[0,1] \longrightarrow L^1[0,1] \ , \text{ with } \frac{1}{p} + \frac{1}{q} = 1 \ ,$$

$\gamma_8(S,H) = S \cdot H$. γ_8 is bilinear and continuous, by the duality of L^p and L^q , and hence is Hadamard differentiable.

$$\gamma_9 : L^1[-m,m] \times \mathbb{R} , \qquad \gamma_9(S) = \int_{-m}^{m} S(t)dt .$$

γ_9 is linear and continuous, and hence Hadamard differentiable.

Now we can write

$$\Phi(G,\theta) = \theta + \int_{-m}^{m} \frac{\lambda A(G(F(z)))M_1(G,\theta)(z)}{\lambda A(G(F(z)))+1-\lambda} + \text{similar term}$$

$$= \theta + \gamma_9 \circ \gamma_8 (\gamma_8(\lambda \gamma_2 \circ \gamma_1(G), M_1(G,\theta)), \gamma_3(\lambda \gamma_2 \circ \gamma_1(G)+(1-\lambda)))$$

$$+ \text{similar term}$$

where $M_1(G,\theta)$ is as in (7.11). Thus it follows that Φ is Hadamard differentiable at (U,θ_o) .

To apply the implicit function theorem, Theorem 6.2.1, it remains to show that conditions (i) and (ii) of that theorem hold for Φ . To prove condition (i) it suffices to show that in a neighborhood of (U,θ_o) ,

$$0 < \alpha \leq \frac{\partial}{\partial \theta} \Phi(G,\theta) \leq \beta < \infty ,$$

for some positive constants α and β . Since the function A is positive valued, and since under some regularity conditions, $\frac{\partial}{\partial \theta} \Phi(G,\theta)$ is continuous near (U,θ_o) , it suffices to prove that

$$(7.12) \qquad \frac{\partial}{\partial \theta} M_i(G,\theta)\Big|_{(U,\theta_o)} > 0 , \qquad i = 1,2 .$$

Let us assume that the d.f.'s F , F_1 and F_2 , are all symmetric about $\theta_o = 0$, then we have

$$(7.13) \qquad \frac{\partial}{\partial \theta} M_i(G,\theta)\bigg|_{(U,0)} = - \frac{\displaystyle\int_0^{F(z)} w_i'(F^{-1}(x))dx}{\displaystyle\int_0^1 w_i(F^{-1}(x))dx} \ .$$

Now w_i is positive and symmetric, so the denominator of (7.13) is positive. We have w_i' odd, and we shall assume that $w_i'(x) \leq 0$ for positive x , which is the case for most common distributions (e.g. normal, Cauchy, logistic, etc.). Under this assumption (7.12) holds for $\theta_o = 0$, and we have

$$0 < \alpha < \frac{\partial}{\partial \theta} \Phi(G,\theta)$$

for (G,θ) near $(U,0)$. The inequality for β will follow simply because all the functions involved have been truncated, and also for this reason condition (ii) of Theorem 6.2.1 will hold. $\qquad \Delta$

CHAPTER VIII

ASYMPTOTIC EFFICIENCY

In this chapter we show that Hadamard differentiability can be used to prove asymptotic efficiency for statistical functionals. Huber (1977) gave a proof that Fréchet differentiable functionals are asymptotically efficient if and only if the influence curve satisfies certain conditions. However he also noted that "the rather stringent regularity conditions - Fréchet differentiability - will rarely be satisfied". Here we show that Huber's result holds under the weaker assumption of Hadamard differentiability. Since we have shown that several classes of statistical functionals are Hadamard differentiable, this approach to asymptotic efficiency through Hadamard differentiability has wide applicability.

Throughout this chapter, we shall consider a parametric family of d.f.'s $F = \{F_\theta : \theta \in \Theta\}$, where Θ is a subset of $\mathbb{R}$.

8.1 Asymptotic efficiency and Hadamard differentiability

Given a parametric family of d.f.'s $F = \{F : \theta \in \Theta\}$, the _Fisher information function_ is defined by

$$I(F_\theta) = \int (\frac{\partial}{\partial\theta} \log f_\theta)^2 dF_\theta \ ,$$

where $F'_\theta = f_\theta$. Note that the existence of $I(F_\theta)$ imposes certain regularity conditions on f_θ .

A statistical functional T whose domain contains a parametric family $F = \{F_\theta : \theta \in \Theta\}$ is said to be a <u>Fisher consistent</u> estimator of θ if $T(F_\theta) = \theta$ for all $\theta \in \Theta$. We say that T is asymptotically efficient when the asymptotic variance of $\sqrt{n}(T(F_n) - T(F_\theta))$ reaches the lower bound $1/I(F_\theta)$.

Suppose that the domain of T has a metric d such that $d(F_\theta, F_{\theta+\delta}) = O(\delta)$ as $\delta \longrightarrow 0$. Following Huber (1977, 1981), we shall say that T is Fréchet differentiable at F_{θ_o}, where θ_o is fixed, if there exists a linear functional T'_{θ_o} defined on the domain of T such that

$$\lim_{\delta \to 0} \frac{T(F_{\theta_o+\delta}) - T(F_{\theta_o}) - T'_{\theta_o}(F_{\theta_o+\delta} - F_{\theta_o})}{d(F_{\theta_o}, F_{\theta_o+\delta})} = 0$$

Using this form of Fréchet derivative, Huber (1977, 1981) proved

<u>Proposition 8.1.1.</u> Let $F = \{F : \theta \in \Theta\}$ be a parametric family of d.f.'s with densities $f = F'_\theta$, and suppose that T is a Fisher consistent estimator of θ. If T is Fréchet differentiable at $F = F_{\theta_o}$ for some $\theta_o \in \Theta$, and if

$$(8.1) \qquad \frac{f_{\theta_o+\delta} - f_{\theta_o}}{\delta \cdot f_{\theta_o}} \xrightarrow{\ L^2(F)\ } \left. \frac{\partial}{\partial \theta} \log f_\theta \right|_{\theta=\theta_o},$$

as $\delta \longrightarrow 0$, and

$$(8.2) \qquad\qquad 0 < I(F) < \infty,$$

then T is asymptotically efficient if and only if

$$(8.3) \qquad IC(x;F,T) = \frac{1}{I(F)} \cdot \left. \frac{\partial}{\partial \theta}(\log f_\theta) \right|_{\theta=\theta_o}.$$

<u>Proof</u>: Since T is Fréchet differentiable at F ,

$$(8.4) \qquad T(F_{\theta_0+\delta}) - T(F) - \int IC(x;F,T)(f_{\theta_0+\delta}(x)-f_{\theta_0}(x))dx$$

$$= o(d(F,F_{\theta_0+\delta}))$$

$$= o(\delta) \ .$$

Since $T(F_{\theta_0+\delta})-T(F) = \delta$, (8.4) and (8.1) imply that

$$\int IC(x;F,T)(\frac{\partial}{\partial\theta} \log f_\theta)(x)\Big|_{\theta=\theta_0} dF(x)$$

$$= \lim_{\delta\to 0} \int IC(x;F,T)(\frac{f_{\theta_0+\delta}(x)-f_{\theta_0}(x)}{\delta\cdot f_{\theta_0}(x)})f_{\theta_0}(x)dx$$

$$= 1 \ .$$

By Schwarz's inequality,

$$\int (IC(x;F,T))^2 dF(x) \geq \frac{1}{I(F)} \ ,$$

with equality if and only if

$$IC(x;F,T) = \frac{1}{I(F)} \cdot \frac{\partial}{\partial\theta} (\log f_\theta)(x)\Big|_{\theta=\theta_0} \ . \qquad\qquad \Delta$$

In the above proposition, it is clear that the Fréchet differentiability of T can be replaced by condition (8.4), which we shall show follows from the Hadamard differentiability of T . To use Hadamard differentiability, we must first extend T to a topological vector space containing F .

Let $\overline{\mathbb{R}} = \mathbb{R} \cup \{+\infty,-\infty\}$ with neighborhoods of $+\infty$ and $-\infty$ of the form $\{x \in \overline{\mathbb{R}}: x > M\}$ and $\{x \in \overline{\mathbb{R}}: x < M\}$, respectively, for $M \neq \pm\infty$. Let $C(\overline{\mathbb{R}})$ be the space of continuous real valued functions on $\mathbb{R}$ such that the limits at $+\infty$ and $-\infty$ exist and let $C(\overline{\mathbb{R}})$ have the uniform topology, so $C(\overline{R})$ is a Banach space.

Suppose that F is a continuous d.f. which is strictly increasing on $\mathbb{R}$. Then F can be extended to a homeomorphism, also denoted by F , $F : \overline{\mathbb{R}} \longrightarrow [0,1]$ with $F(-\infty) = 0$ and $F(+\infty) = 1$. The inverse of F , F^{-1}, is also a homeomorphism $F^{-1} : [0,1] \longrightarrow \overline{\mathbb{R}}$. This can be used to induce a norm perserving transformation from $C(\overline{\mathbb{R}})$ to $C[0,1]$ defined by $G \longrightarrow G \circ F^{-1}$, where $G \in C(\overline{\mathbb{R}})$ and F is fixed.

<u>Lemma 8.1.2</u>. Let F be a parametric family of d.f.'s and let T be a statistical functional. Suppose that for some θ_o , $F = F_{\theta_o} \in F$ is continuous and strictly increasing on $\mathbb{R}$. If the induced functional τ , such that $\tau(F_\theta \circ F^{-1}) = T(F_\theta)$, $F_\theta \in F$, can be extended to $C[0,1]$, then T can be extended to $C(\overline{\mathbb{R}})$. If τ is Hadamard differentiable at $U \in C[0,1]$ (where $U(x) = x$) then T is Hadamard differentiable at $F \in C(\overline{\mathbb{R}})$.

<u>Proof</u>: Let $G \in C(\overline{\mathbb{R}})$. Then $G \circ F^{-1} \in C[0,1]$, so if τ is defined on $C[0,1]$, we can extend T to $C(\overline{\mathbb{R}})$ by defining

$$T(G) = \tau(G \circ F^{-1}) \ .$$

Then $T : C(\overline{\mathbb{R}}) \longrightarrow \mathbb{R}$ with $T(F) = \tau(U)$.

Suppose that τ is Hadamard differentiable at U . The mapping $G \longrightarrow G \circ F^{-1}$ from $C(\overline{\mathbb{R}})$ to $C[0,1]$ is linear and continuous and therefore Hadamard differentiable. Hence T is the composition of two Hadamard differentiable functions and it is also Hadamard differentiable at $F = C(\overline{\mathbb{R}})$. Δ

In the above proof, since the transformation $G \longrightarrow G \circ F^{-1}$ is norm preserving, it follows that the compact sets in $C(\overline{\mathbb{R}})$ are precisely those sets whose images are compact in $C[0,1]$.

Now that we have shown how to extend T to the normed vector space $C(\overline{\mathbb{R}})$, we shall prove that the weaker requirement of Hadamard differentiability can replace Fréchet differentiability to obtain (8.3) of Proposition 8.1.1.

<u>Theorem 8.1.3.</u> Let $F = \{F_\theta : \theta \in \Theta\}$ be a parametric family of d.f.'s with densities $f_\theta = F'_\theta$, and suppose that the mapping $\theta \longrightarrow F_\theta$ from Θ to $C(\overline{\mathbb{R}})$ is continuous. Let T be a statistical functional which is a Fisher consistent estimator of θ , and assume that T can be extended to $C(\overline{\mathbb{R}})$. If T is Hadamard differentiable at $F = F_{\theta_0}$ for some $\theta_0 \in \Theta$, where F is strictly increasing, and if (8.1) and (8.2) hold, then T is asymptotically efficient if and only if (8.3) is satisfied.

<u>Proof</u>: As we noted above, it is sufficient to prove that

$$(8.5) \qquad T(F_{\theta_0+\delta}) - T(F) - \int IC(x;F,T)d(F_{\theta_0+\delta}-F)(x) = o(\delta) \ ,$$

as was done in the proof of Proposition 8.1.1.

Since T is Hadamard differentiable at $F = F_{\theta_0}$, the derivative T'_F exists and

$$(8.6) \qquad \frac{T(F_{\theta_0+\delta}) - T(F) - T'_F(F_{\theta_0+\delta}-F)}{\delta}$$

$$= \frac{T(F+\delta(F_{\theta_0+\delta}-F)/\delta) - T(F) - T'_F(\delta(F_{\theta_0+\delta}-F)/\delta)}{\delta} \ .$$

Let

$$g(x) = \int_{-\infty}^{x} \frac{\partial}{\partial\theta} \log f_\theta \Big|_{\theta=\theta_0} dF(t) \ .$$

By condition (8.2), g is well defined and $g \in C(\overline{\mathbb{R}})$. Now, define

$$(8.7) \qquad H_\delta = \begin{cases} (F_{\theta_o + \delta} - F)/\delta & \text{if } \delta \neq 0 \\[2em] g & \text{if } \delta = 0 \ . \end{cases}$$

Then (8.6) can be written

$$(8.8) \qquad \frac{T(F + \delta H_\delta) - T(F) - T'_F(\delta H_\delta)}{\delta} \ .$$

Consider the set $K = \{H_\delta : \delta \in [-1,1]\}$. If we can show that $K \subset C(\overline{\mathbb{R}})$ is compact, then, since T is Hadamard differentiable at F , (8.8) tends to zero as $\delta \longrightarrow 0$ and the theorem is proved. Therefore we have only to show that K is compact in $C(\overline{\mathbb{R}})$.

Consider the mapping $\phi : [-1,1] \longrightarrow C(\mathbb{R})$ defined by $\phi(\delta) = H_\delta$. The image of $[-1,1]$ by ϕ is the set K , so if we can show that ϕ is continuous, it will follow that K is compact.

For $\delta \neq 0$ the numerator and denominator of H_δ in (8.7) are both continuous in δ since by hypothesis the mapping $\theta \longrightarrow F_\theta$ is continuous. Therefore ϕ is continuous at all $\delta \neq 0$.

To prove the continuity of ϕ at zero we must show that for any $\varepsilon > 0$,

$$\|H_\delta - g\| < \varepsilon$$

for δ small enough, since $g = H_o$.

Now,

$$\|H_\delta - g\| = \sup_{x \in \mathbb{R}} \left| \frac{F_{\theta_o + \delta}(x) - F(x)}{\delta} - \int_{-\infty}^{x} \frac{\partial}{\partial \theta} \log f_\theta(t) \Big|_{\theta = \theta_o} dF(t) \right|$$

$$(8.9) \qquad = \sup_{x \in \mathbb{R}} \left| \int_{-\infty}^{x} \left[\frac{f_{\theta_o + \delta}(t) - f_{\theta_o}(t)}{\delta f_{\theta_o}(t)} - \frac{\partial}{\partial \theta} \log f_\theta(t) \Big|_{\theta = \theta_o} \right] dF(t) \right|$$

$$= \sup_{x \in \mathbb{R}} \left| \int I_{(-\infty,x)}(t) \left[\frac{f_{\theta_o+\delta}(t)-f_{\theta_o}(t)}{\delta f_{\theta_o}(t)} - \frac{\partial}{\partial\theta} \log f_\theta(t) \Big|_{\theta=\theta_o} \right] dF(t) \right|$$

$$\leq \sup_{x \in \mathbb{R}} \left\| I_{(-\infty,x)} \right\|_{L^2(F)} \left\| \frac{f_{\theta_o+\delta}-f_{\theta_o}}{\delta f_{\theta_o}} - \frac{\partial}{\partial\theta} \log f_\theta \Big|_{\theta=\theta_o} \right\|_{L^2(F)}$$

by the Cauchy-Schwarz inequality.

But $\quad \sup\limits_{x \in \mathbb{R}} \left\| I_{(-\infty,x)} \right\|_{L^2(F)} = 1 \quad$ and

$$\left\| \frac{f_{\theta_o+\delta}-f_{\theta_o}}{\delta f_{\theta_o}} - \frac{\partial}{\partial\theta} \log f_\theta \Big|_{\theta=\theta_o} \right\|_{L^2(F)} \longrightarrow 0$$

as $\delta \longrightarrow 0$ by condition (8.1). Therefore (8.9) tends to zero, so ϕ is continuous at zero and the theorem is proved. $\qquad \Delta$

8.2 Asymptotically efficient estimators of location

We are interested in applying the results of the previous section to a parametric family $F = \{F_\theta : \theta \in \Theta\}$ where θ is a location parameter. In this case we can assume without loss of generality that $\theta_o = 0$. Let $f_\theta = F'_\theta$. Then we have

$$\frac{\partial}{\partial\theta} \log f_\theta \Big|_{\theta=\theta_o} = -f'_0/f_0 .$$

Theorem 8.2.1. Let F be a parametric family of absolutely continuous d.f.'s with location parameter θ. Let T be a statistical functional whose domain contains F with $T(F_\theta) = \theta$ and suppose that $F = F_0$ is strictly increasing. Assume that the functional τ defined such that $\tau(F_\theta \circ F^{-1}) = T(F_\theta)$ can be extended to $C[0,1]$ and is Hadamard differentiable at U. If

$$(8.10) \qquad \frac{f_\delta - f}{\delta f} \xrightarrow{\ L^2(F)\ } -f'/f$$

and $0 < I(F) < \infty$, then T is asymptotically efficient if and only if

$$(8.11) \qquad IC(x;F,T) = \frac{-f'(x)}{I(F)f(x)} \ .$$

<u>Proof</u>: It suffices to show the hypotheses of Theorem 8.1.3 are satisfied. This amounts to proving that the mapping $\theta \longrightarrow F_\theta$ is continuous in θ , which follows immediately since $F_\theta(x) = F(x-\theta)$ and F is absolutely continuous. $\qquad\qquad\qquad\qquad\qquad\qquad\qquad\qquad\qquad \Delta$

Condition (8.10) above will usually be satisfied for reasonable classes of d.f.'s, so condition (8.11) becomes the critical factor in determining whether or not a statistical functional is asymptotically efficient. Let us see what this means in some specific cases. For more details, see Huber (1981).

<u>Example 8.2.2</u>. For M-estimators as defined in section 5.1, condition (8.11) holds when

$$\psi(x) = -c \ f'(x)/f(x)$$

with

$$c = \int \psi'(x)\,dF(x)/I(F) \ . \qquad\qquad\qquad\qquad\qquad \Delta$$

<u>Example 8.2.3</u>. For L-estimators as defined in section 5.2, when $h(x) = x$ condition (8.11) is satisfied if m is a measure on $(0,1)$ such that

$$m(F(x)) = - \frac{d}{dx} \ (f'(x)/f(x))/I(F) \ . \qquad\qquad\qquad \Delta$$

<u>Example 8.2.4</u>. For R-estimators as defined in section 5.3, for a symmetric d.f. F condition (8.11) holds when

$$J(2F(x)-1) = -c \ f'(x)/f(x)$$

with

$$c = 2\int J'(2F(x)-1)f^2(x)\,dx/I(F) \ . \qquad\qquad \Delta$$

<u>Example 8.2.5</u>. For the sample median, (8.11) holds when

$$f'(x)/f(x) = \begin{cases} k(2f(0))^{-1} & \text{if} \ \ x \geq 0 \\[3mm] -k(2f(0))^{-1} & \text{if} \ \ x < 0 \end{cases}$$

where $k < 0$. This corresponds to an exponential distribution. $\qquad \Delta$

REFERENCES

ANDREWS, D.F. et al. (1972). Robust Estimates of Location. Princeton
        University Press, Princeton.

AVERBUKH, V.I. and SMOLYANOV, O.G. (1968). The various definitions of
        the derivative in linear topological spaces. Russian Math.
        Surveys 23, 67-113.

BERK, R.H. (1977), Unpublished lecture notes, Statistics Dept.,
        Rutgers University, New Brunswick, N.J.

BERAN, R. (1977). Robust location estimates. Ann. Statist. 5, 431-444.

BILLINGSLEY, P. (1968). Convergence of Probability Measures, John Wiley
        & Sons, New York.

BOOS, D.D. (1979). A differential for L-statistics. Ann. Statist. 7,
        955-959.

BOOS, D.D. and SERFLING R.J. (1980). A note on differentials and
        the CLT and LIL for statistical functions with applications
        to M-estimates. Ann. Statist. 8, 618-624.

DIEUDONNE, J. (1960). Foundations of Modern Analysis. Academic Press,
        New York.

DONSKER, M. (1952). Justification and extension of Doob's heuristic
        approach to the Kolmogorov-Smirnov theorems. Ann. Math.
        Statist. 23, 277-281.

DOOB, J.L. (1949). Heuristic approach to the Kolmogorov-Smirnov
        theorems. Ann. Math. Statist. 20, 393-403.

DUNFORD, N. and SCHWARTZ, J.T. (1958). Linear Operators, Part I.
        Interscience, New York.

FERNHOLZ, L.T. (1979). Topics in Mathematical Statistics and Probability
        Theory. Ph.D. dissertation, Rutgers University, New
        Brunswick, N.J.

FILIPPOVA, A.A. (1962). Mises theorem on the asymptotic behavior of
        functionals of empirical distribution function and its
        statistical applications. Theory Prob. Appl. 7, 24-57.

GUARINO, R. (1981). Robust estimation under a finite number of alterna-
        tives: compromise estimates of location. Ph.D. Disserta-
        tion, Princeton University, Princeton, N.J.

HAMPEL, F.R. (1968). Contributions to the theory of robust estimation. Ph.D. dissertation, Univ. of Calif., Berkeley, CA.

———————— (1971). A general qualitative definition of robustness. Ann. Math. Statist. 42, 1887-1896.

———————— (1974). The influence curve and its role in robust estimation. J. Amer. Statist. Assoc. 69, 383-393.

HODGES, J.L. and LEHMANN, E.L. (1963). Estimates of location based on rank tests. Ann. Math. Statist. 34, 598-611.

HUBER, P.J. (1964). Robust estimation of a location parameter. Ann. Math. Statist. 35, 73-101.

———————— (1977). Robust Statistical Procedures, Regional Conference Series in Applied Mathematics 27, SIAM.

———————— (1981). Robust Statistics. John Wiley & Sons, New York.

KALLIANPUR, G. and RAO, C.R. (1955). On Fisher's lower bound to asymptotic variance of a consistent estimate. Sankhya 15, 331-342.

KRAFT, C. (1955). Some conditions for consistency and uniform consistency of statistical procedures. Univ. of Calif. Publ. in Statistics, Vol. 2, No. 6, 125-142.

KELLER, H.H. (1974). Differential Calculus in Locally Convex Spaces, Lecture Notes in Math., No. 417, Springer-Verlag, Berlin.

VON MISES, R. (1936). Les lois de probabilite pour les fonctions statistiques. Ann. Inst. H. Poincaré B., 6, 185-212.

———————— (1937). Sur les fonctions statistiques. Soc. Math. de France, Conference de la Reunion Internat. des Math., Paris.

———————— (1947). On the asymptotic distribution of differentiable statistical functions. Ann. Math. Statist. 18, 309-348.

REEDS, J.A. (1976). On the definition of von Mises functionals, Ph.D. dissertation, Harvard Univ., Cambridge, MA.

SERFLING, R.J. (1980). Approximation Theorems of Mathematical Statistics. John Wiley & Sons, New York.

YAMAMURO, S. (1974). Differential Calculus in Topological Linear Spaces. Lecture Notes in Math., No. 374, Springer-Verlag, Berlin.

<u>LIST OF SYMBOLS AND ABBREVIATIONS</u>

i.i.d.   independent identically distributed

d.f.   distribution function

a.s.   almost surely

a.e.   almost everywhere

$\xrightarrow{P}$ convergence in probability

$\xrightarrow{\mathcal{D}}$ convergence in distribution

$\xrightarrow{L^p}$ convergence in L^p or in p-th mean.

$o(t)$ $f(t) = o(t)$ if $\lim\limits_{t \to a} f(t)/t = 0$, usually $a = 0$ or $a = \pm\infty$.

$o_P(a_n)$ $X_n = o_P(a_n)$ if $X_n/a_n \xrightarrow{P} 0$ as $n \longrightarrow \infty$.

$O_P(a_n)$ $X_n = O_P(a_n)$ if $\forall \varepsilon > 0$, there exist constants N_ε and K_ε such that $P\{|X_n/a_n| > K_\varepsilon\} \le \varepsilon$ for $n > N_\varepsilon$.

$\mathbb{R}$ the set of real numbers

$N(\mu,\sigma^2)$ the normal or gaussian distribution with mean μ and variance σ^2 .

Δ indicates the end of a theorem or example.

Lecture Notes in Statistics

(Continued from page ii)

Vol. 18: W. Britton, Conjugate Duality and the Exponential Fourier Spectrum. v, 226 pages, 1983.

Vol. 19: L. Fernholz, von Mises Calculus For Statistical Functionals. viii, 124 pages, 1983.